# A Fynbos Year

*Protea scolymocephala*

# A Fynbos Year

illustrated by **Liz McMahon**

with text by **Michael Fraser**

Cape Town & Johannesburg **dp** *DAVID PHILIP*

First published 1988 by David Philip, Publisher (Pty) Ltd,
217 Werdmuller Centre, Claremont 7700, South Africa

ISBN 0–86486–103–6 (cased)    ISBN 0–86486–123–0 (paper)

*To our parents*

Printed by National Book Printers, Zelda Street, Goodwood,
Cape, South Africa

# Contents

*Leucadendron*

# Preface

On arriving at the Cape in early 1984, our first impression, that fynbos looked much like the Scottish moorlands we had lately abandoned, did not last long. Although the rugged scenery and short, windcropped vegetation of Cape Point (where we were to be based for two years) looked uncomfortably familiar, no Scottish glen ever echoed to the bark of baboons or basked in such brilliant sunshine. On the day our motorbike was overtaken by a speeding Ostrich any remaining doubts were dispelled. This was definitely not Scotland. And, proud as we are of our famous heather-clad hillsides, it was humbling to discover that in answer to our two or three, the southern Cape boasts well over 600 species of *Erica* (heaths). This certainly put fynbos into perspective.

An illustrated account of the plants and animals we encountered seemed to us the natural product of our experiences and accumulation of scribbled field-notes and rough sketches. To describe and illustrate everything we came across (itself a tiny fraction of what fynbos has to offer) would be the work of many lifetimes. A compromise, acceptable we hope, has been to portray a sample of the 'everyday' plants and animals which we found, and any other walker, birder or botanist may find, over the course of one year, and add to it a review of fynbos natural history. In this way, the illustrations can be seen as representatives of a very much wider, richer system. If birds have remained our primary interest it is not, we hope, at the expense of the amazing wealth of other fynbos inhabitants.

*A Fynbos Year* is, therefore, a general account of fynbos illustrated with characteristic plants and animals as we found them between February 1987 and January 1988. As such, it represents our plea for the conservation of this fascinating, beautiful and highly endangered floral kingdom.

Our association with fynbos began when M.F. was a member of the Percy FitzPatrick Institute of African Ornithology (PFIAO) at the University of Cape Town (UCT) and funded by a J. W. Jagger Overseas Student Scholarship, through UCT, and by the Nature Conservation Section (Invasive Biota) and Terrestrial Ecosystems Section (Fynbos Biome Project) of the Council for Scientific and Industrial Research's National Programme for Ecosystem Research.

During our 'fynbos year', access to areas under their control was granted by Chief Warden Gerald Wright (Cape of Good Hope Nature Reserve), the Forestry Branch of the Department of Environment Affairs (Jonkershoek State Forest), Nico Myburgh (Helderberg Nature Reserve) and Mr and Mrs Alastair Trafford (*Mont Fleur*).

A licence to collect plants was granted by the Cape Department of Nature and Environment Conservation. Plant and animal specimens were identified or provided by Gerald Wright; Atherton de Villiers and Stewart Thorne (Chief Directorate, Department of Nature and Environmental Conservation, Jonkershoek); Mark and Zelda Wright, Werner Techman and Dr Mike Jarvis (Department of Agriculture, Elsenburg); Pat Brown, Greg Forsyth, Anneke de Kock and Dave le Maitre (Jonkershoek Forestry Research Centre (JFRC)); Dr Gary Williams (South African Museum); Howard Langley and Gill Wheeler (Rondevlei Bird Sanctuary); Prof. Jenny Jarvis, Dr Mike Picker, Jean Harris and Dr Barry Lovegrove (Department of Zoology, UCT); Kobus Steenkamp (South African Nature Foundation, *Protea Heights*); Dave Pepler (Department of Nature Conservation, University of Stellenbosch); Dr Peter Linder (Bolus Herbarium, UCT); Sally Klute (Somerset West) and David Allan (PFIAO).

Peter Ryan (PFIAO), Dr Brian van Wilgen and Dave Richardson (JFRC), Rob Martin (Faculty of Forestry, University of Stellenbosch), Barry Watkins (PFIAO), Dave Pepler, Atherton de Villiers and Stewart Thorne reviewed the text wholly or in part. Any remaining inaccuracies therein are our own.

David Philip and Russell Martin of David Philip, Publisher supported the project from the outset and provided much help and advice.

To all these individuals and institutions we extend our sincere thanks.

L.M. expresses her gratitude to Gerry Keenan, Stan Clement–Smith and Ron Stenberg, whose artistic guidance, encouragement and enthusiasm led her into a career as a freelance illustrator.

M.F. would like to thank Miss Elizabeth Fraser for her generous support. The guidance and friendship over many years of his ornithological mentor and bird-ringing trainer, Ian Balfour–Paul, are in no small way responsible for M.F.'s preoccupation with things feathered.

Finally, we give our heartfelt thanks to our parents and families, who have not only tolerated our often eccentric pursuits, but actively encouraged them.

Liz McMahon & Michael Fraser
August 1988

# Introduction

## The smallest kingdom

'Fynbos' (pronounced *fayn-bos*) is the Afrikaans word meaning 'fine bush', from the Dutch *fijn bosch*. It describes the narrow-leaved plants which characterise much of the vegetation of the southern and southwestern Cape Province, South Africa. Fynbos contains many broad-leaved, as well as fine-leaved plants, however, and is made up of a number of distinct vegetation types. But the name has persisted and has become synonymous with one of the richest and, perhaps, most beautiful floras in the world.

In its broadest sense, fynbos comprises over twenty related vegetation categories. Together, these are known as the Fynbos Biome, a 'biome' being a collection of vegetation formations sharing certain environmental features, notably similar structure. An exceptionally large number of plant species, especially those found only within its bounds and nowhere else (so-called 'endemic' species), has earned the Fynbos Biome international recognition as one of the world's floral kingdoms. Such an accolade is awarded only to those regions which support a characteristic flora of their own that has been conditioned by local history and circumstances. There are five other such plant kingdoms, each with unique botanical attributes, but the relative size of fynbos – the Cape Floral Kingdom – makes it particularly remarkable. The Australian Kingdom is just that – the whole of Australia. The Boreal Kingdom extends over most of the northern hemisphere and occupies more than 40 per cent of the world's land surface. The Cape Floral Kingdom, perched on the southern rim of the African continent, is one-thousandth the size of the Boreal Kingdom and occupies only 0,04 per cent of the world's land surface.

The Fynbos Biome (some 70 000km² in extent) stretches in a narrow, crescent-shaped arc, from a northern limit on the Nieuwoudtville Escarpment, 350km south to the Cape Peninsula, then almost 750km east along the southern Cape coast as far as Port Elizabeth and inland to Grahamstown, where only scattered fragments occur. Nowhere along its length does fynbos extend more than 200km from the sea (the cold Atlantic Ocean in the west and the much warmer Indian Ocean in the south), and in places it may be as little as 40km wide.

*Mimetes cucullatus*

*Sirkelsvlei, a typical Mountain Fynbos lakelet at the Cape of Good Hope Nature Reserve*

*Mountain Fynbos vegetation amongst exposed quartzitic sandstone at the Cape of Good Hope Reserve*

# The illustrations

The natural history sketchbook is not a recent phenomenon. Early explorers and naturalists who undertook journeys of discovery often combined scientific expertise with the skill of the artist. In the days before photography, the only way to record what they found was to press it, bottle it, stuff it or paint it. However, the advent of the camera did not see the decline of natural history illustration, and wildlife painting has remained an art in its own right.

British painters such as Charles Tunnicliffe, Eric Ennion, Peter Scott, Donald Watson and others belong to a long tradition of artist–naturalists. Heinie von Michaelis is a notable South African member of this genre. More recently, Janet Marsh, Lars Jonsson and Keith Brockie are artists of a new generation who continue this tradition.

Although the idea of *A Fynbos Year* was an almost immediate response to the wealth of subjects that fynbos has to offer, it was not until February 1987 that painting for the book began.

There is no substitute for being out in the field, with or without sketching materials, gaining first-hand experience of the subjects to be painted. Thankfully, plants stay in one spot, but a telescope and binoculars are essential for observing those mammals and birds which are not so approachable. Birds caught for ringing can be held briefly, providing the opportunity for making quick notes of plumage and other characteristics. The experience of handling the bird itself gives much insight into the idiosyncrasies of individual species. Where possible, use was made of tame animals when sightings in the field were fleeting. Insects were also drawn largely 'in captivity', housed temporarily in clear containers before being released.

Most of the groundwork for the plates was done in the field, whether in the form of detailed studies of plants or a collection of quick scribbles of animals on the move. However, the paintings were generally completed on the kitchen table! The dates on the plates refer to the day when the subject was found and not necessarily to when the illustration was completed. Watercolour paints and coloured pencils are ideal for quick field sketches and for flower painting. On coloured paper gouache was preferred, however, allowing the paper to do some of the work.

In the hunt for subjects, a combination of necessity and choice has restricted us to areas of the southwest Cape where our research has been carried out. The three sites that feature most prominently are, therefore, the Cape of Good Hope Nature Reserve at the southern tip of the Cape Peninsula, the Jonkershoek Valley (in particular the Swartboskloof subcatchment) near Stellenbosch, and Helderberg Nature Reserve at Somerset West. Although these could have been selected for our convenience alone, they incorporate fine tracts of fynbos vegetation and are well known to walkers, naturalists and sightseers. For many people, a visit to one of these places may provide their introduction to fynbos. To choose three more beautiful and impressive areas in which to work would certainly have been difficult.

## Cape of Good Hope Nature Reserve

The Cape of Good Hope Nature Reserve (also known as Cape Point) is scenically magnificent and one of South Africa's most historical landmarks. In 1488, the Portuguese navigator Bartholomeu Dias rounded this southwest tip of Africa and named it the Cape of Storms. Sir Francis Drake described the distinctive headland as 'the fairest and most stately thing we saw in the whole circumference of the Globe' when he rounded the Cape on 18 June 1580. Drake obviously hit better weather than the unfortunate Dias, who was wrecked and drowned on a later voyage on this route.

The reserve is 7 750ha in extent and is composed of level or gently inclined beds of Table Mountain Sandstone resting on Cape Granite. It is bounded by the warm waters of False Bay to the east and the much colder upwellings of the Benguela current in the Atlantic on the west. In summer, persistent southeast winds buffet the Peninsula and may be felt most strongly at the famous landmarks of Cape Point and the Cape of Good Hope. These winds have a cooling effect so that there is little difference between average summer and winter temperatures (18° and 13°, respectively). Some 70 per cent of rainfall occurs in the winter months (June–August), and ranges from 355mm a year at the Cape Point lighthouse to 698mm at Smitswinkel Bay in the northeast corner of the reserve.

The vegetation comprises Mountain Fynbos and a narrow coastal strip of Strandveld. Of the almost 1 100 plant species, twelve, including five *Ericas*, are endemic to the reserve – that is, they are found nowhere else in the *world*. A further thirty species are rare or endangered. The rich plant communities have survived despite intensive agriculture during the nineteenth and early twentieth centuries, the presence of almost 6 000 horses during the Anglo–Boer War (1899–1902), and an early management regime which attempted to convert the fynbos vegetation, by means of brush-cutting, ploughing, fertilising and sowing with exotic pasture grasses, into a game park with a selection of large, alien mammals introduced for public spectacle. Now there are small numbers of Bontebok, Eland and Red Hartebeest in addition to the naturally occurring best place we know to find that fynbos speciality, the Cape Siskin.

## Jonkershoek Valley and Swartboskloof

The Jonkershoek Valley is situated 10km southeast of the historic country town of Stellenbosch. Its spectacular scenery is dominated by the twin peaks of Jonkershoek, rising to 1 494m on the north flank of the valley through which the Eerste River flows.

The top of the Dwarsberg at the head of the valley holds the dubious distinction of recording the highest annual rainfall in South Africa: 3 620mm. This is even more remarkable in that the annual rainfall in the valley bottom is 1 180mm and that of Stellenbosch only 780mm! On another of the valley's mountain tops, Victoria Peak (1 560m), an automatic weather station has monitored frequent winds of 100k/h or more, with subzero winter temperatures persisting for some days. Under such conditions, it is hardly surprising that the vegetation there displays many alpine characteristics.

The Jonkershoek Valley is the site of a long-term investigation into the effects of afforestation on fynbos stream-flow and erosion, and the management of unafforested areas in the mountain catchments of the western Cape. This and other work is carried out by the Forestry Branch of the Department of Environment Affairs from the Jonkershoek Forestry Research Centre, which in 1987 celebrated 50 years of research.

Although designated a State Forest, most of the Jonkershoek Valley – 10 930ha in extent – consists of natural Mountain Fynbos and indigenous forest patches. Monterey Pines have been planted on 300ha of the 3 000ha that fall within the Eerste River catchment in the lower valley.

Swartboskloof, a small, fan-shaped offshoot on the south side of the valley, has received much attention as one of the primary sites for the study of mountain ecosystems identified by the CSIR's Fynbos Biome Steering Committee. The 375ha kloof was burnt in a controlled fire in March 1987, and on-going research before and after the burn is aimed at elucidating its effects on the fauna and flora, as well as physical aspects such as stream-flow and runoff.

Swartboskloof has an exceptionally rich flora, supporting 651 species of flowering plant and ferns on its granite- and sandstone-derived soils. Tall proteoid shrublands, comprising predominantly Blue Sugarbush, Sugarbush and Waboom, dominate the lower slopes. Above 850m the vegetation is largely replaced by low herbland communities. Mountain Cypress trees are found in scattered stands, and patches and narrow strips of forest occur on boulder screes and fringe the two main perennial streams.

In all, 26 species of mammal have been found at Swartboskloof. Ten of these are mice or shrews, but Leopard, Honey Badger and Antbear also occur.

Mountain Fynbos and forest birds are found in Swartboskloof, 74 species having been recorded in all. All six fynbos endemics occur here, and it is a

*Winter in the Jonkershoek Valley with snow on the mountain peaks*

buck, notably Grey Rhebok and Grysbok. By far the most important Cape Point animal resident is the handsome (or ugly, depending on your taste) and very rare Cape Platanna.

Another and very pernicious threat to the indigenous vegetation is invasion by alien woody plants, notably Australian *Acacias*. To prevent these aliens from taking over entirely, which they would surely do, a major eradication programme has been embarked on by the reserve authorities.

Our work at Cape Point involved an assessment of the effects of aliens and a block-burning regime on the indigenous birds. Many of the over 200 bird species recorded from the reserve are coastal or oceanic, but a variety of bush and montane birds does occur. These range from the ubiquitous Greybacked Cisticola and Bokmakierie, to the much rarer Striped Flufftail and Blackrumped (Hottentot) Buttonquail. The scrubby area around the Cape Point lighthouse buildings is the

*Swartboskloof. Patches of woodland are visible in the centre distance, while the left-hand slope still bears evidence of burning two months previously.*

particularly good place to find Protea Canary. Cape Sugarbirds and Orangebreasted Sunbirds are abundant when their food plants are in flower, but are rare or absent at other times. Paradise Flycatchers occupy the forest in summer months, and Olive Woodpecker, Cape Batis and Southern Boubou are among the species that may be found there at any time of year.

Our work in Swartboskloof involved an assessment of changes in bird populations following the fire. Frequent visits also gave us the opportunity to observe the succession of plants, with different species appearing and flowering at different stages after the fire. The vegetation was 28 years old when it was burnt, and thereafter a number of plant species which had not been recorded over the intervening years emerged from dormant seeds or bulbs. The displays of *Gladiolus, Aristea, Wachendorfia* and *Watsonia*, in particular, were very impressive. Incidentally, the fire appears to have had little effect on the birds, much to our surprise.

*Watsonias flowering in Swartboskloof six months after a fire*

*Helderberg Nature Reserve, Somerset West*

### Helderberg Nature Reserve

We have made at least monthly visits to Helderberg Nature Reserve during our fynbos year. Here the Somerset West Municipality maintains a protected area, formally proclaimed in September 1960, which includes an extensive area of Mountain Fynbos on the hillside, leading down to a garden that is planted with a variety of indigenous species. When in flower, the dense patches of *Ericas*, *Proteas* and pincushions attract hundreds of nectar-feeding birds. Most abundant are Cape Sugarbirds, particularly when the orange pincushion *Leucospermum cordifolium* is in bloom (August–December, but peaking in September–October).

Curious to know where the sugarbirds went to when there were no food plants in flower, we embarked on a programme of bird-ringing in the gardens. The efforts of the Helderberg Ringing Group have been well rewarded with over one thousand Cape Sugarbirds and a further thousand birds of 34 other species now ringed there.

### Elsewhere

*A Fynbos Year* has also taken us to some of the more outlying areas of the biome, east through Knysna to Port Elizabeth along the picturesque Garden Route, and north from Cape Town over spectacular mountain passes to Ceres and on to Karoopoort, where, over the width of a dusty road, fynbos vegetation gives way to the arid and largely desertified Karoo. Closer to home, visits were made to the coastal area around Betty's Bay which is rich in plants but suffers from alien encroachment and road and housing development. Table Mountain, sadly overdeveloped from its lower slopes to the summit plateau, is well worth a visit if only to ponder how the landscape must have looked in times past. The view to the south of Table Mountain is as impressive as that to the north is depressing. Blaauwklippen is one of the most beautiful valleys of the Stellenbosch area, and at its head a tract of unspoilt Mountain Fynbos is maintained by the Trafford family – a fine example of the contribution the private landowner can make to conservation.

# A Fynbos Year

A splash of pink *Tritoniopsis*
amongst the browns and greens
of restios, grasses and the toothbrush fern.

Swartboskloof 13 th January.

2

3

Cape Eagle Owl
Bubo capensis
(Kaapse Ooruil)

Somerset West,
17th January.

4

Spotted Eagle Owl
_Bubo africanus_
(Gevlekte Ooruil)

Somerset West,
18th January.

5

Betty's Bay, 21st January.

The Striped Mouse (Streepmuis)
<u>Rhabdomys pumilio</u> is very
active by day and is one of
the most frequently seen small
mammals in fynbos. It is easily
identified by the stripes down
its back.

6

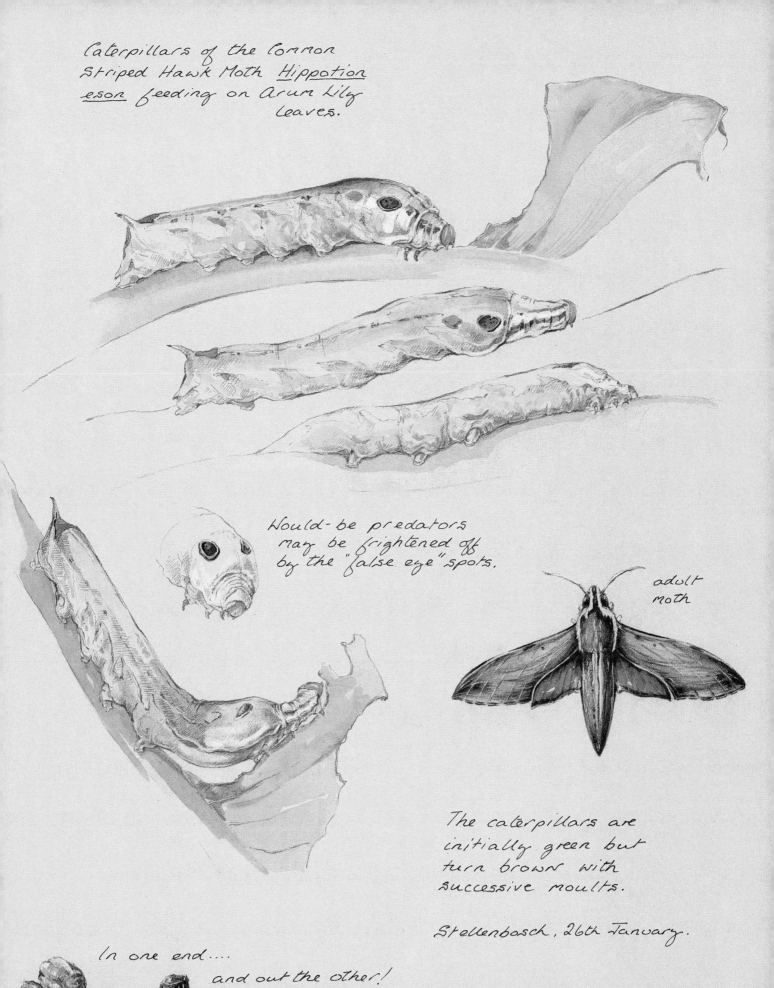

Caterpillars of the Common
Striped Hawk Moth <u>Hippotion</u>
<u>esor</u> feeding on Arum Lily
Leaves.

Would-be predators
may be frightened off
by the "false eye" spots.

adult
moth

The caterpillars are
initially green but
turn brown with
successive moults.

Stellenbosch, 26th January.

In one end....
and out the other!

7

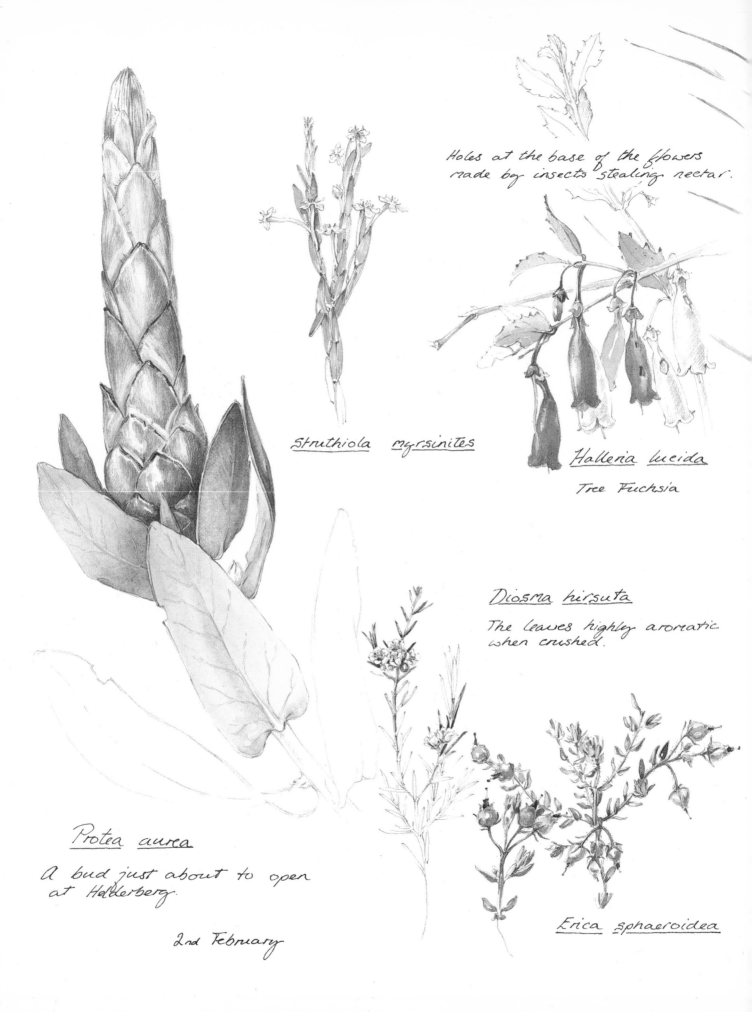

Holes at the base of the flowers
made by insects stealing nectar.

Struthiola myrsinites

Halleria lucida

Tree Fuchsia

Diosma hirsuta

The leaves highly aromatic
when crushed.

Protea aurea

A bud just about to open
at Helderberg.

2nd February

Erica sphaeroidea

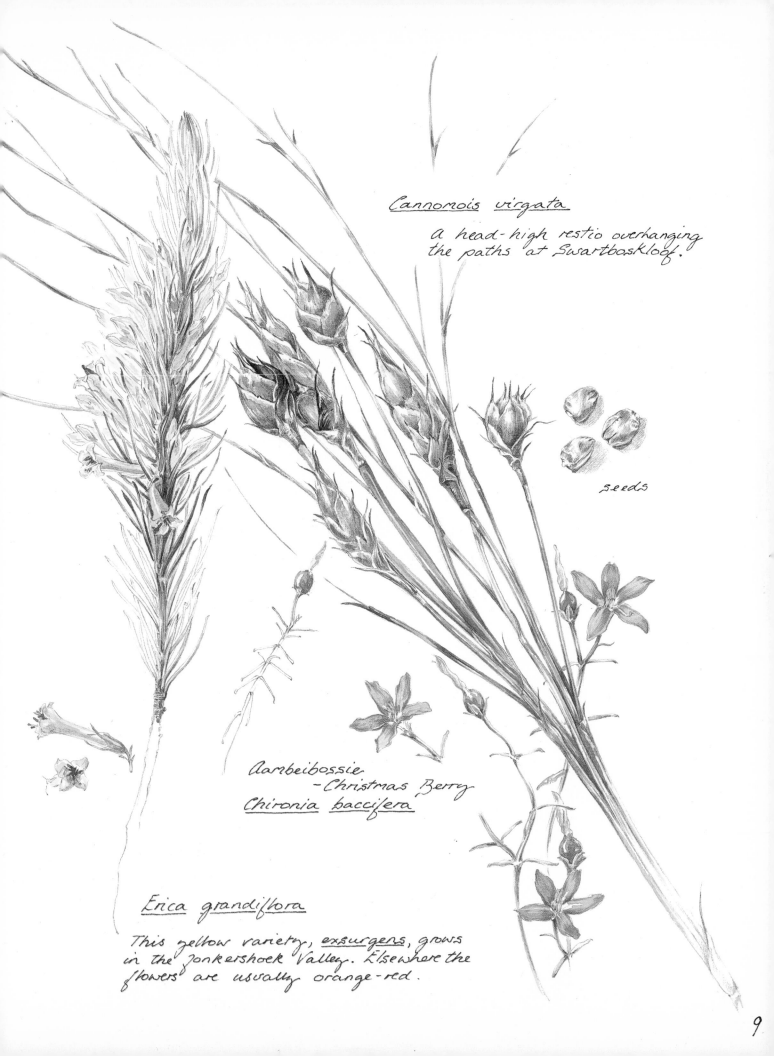

*Cannomois virgata*

A head-high restio overhanging
the paths at Swartboskloof.

seeds

Aambeibossie
— Christmas Berry
*Chironia baccifera*

*Erica grandiflora*

This yellow variety, _exsurgens_, grows
in the Jonkershoek Valley. Elsewhere the
flowers are usually orange-red.

9

Grysbok <u>Raphicerus</u> <u>melanotis</u>
Endemic to fynbos. The grey
grizzled coat distinguishes
this small antelope from the
Steenbok  R. campestris.
Females and youngsters,
like this one lack the
short pointed horns of
the male.

Brightwater,
Cape Point Nature Reserve.
9th February

10

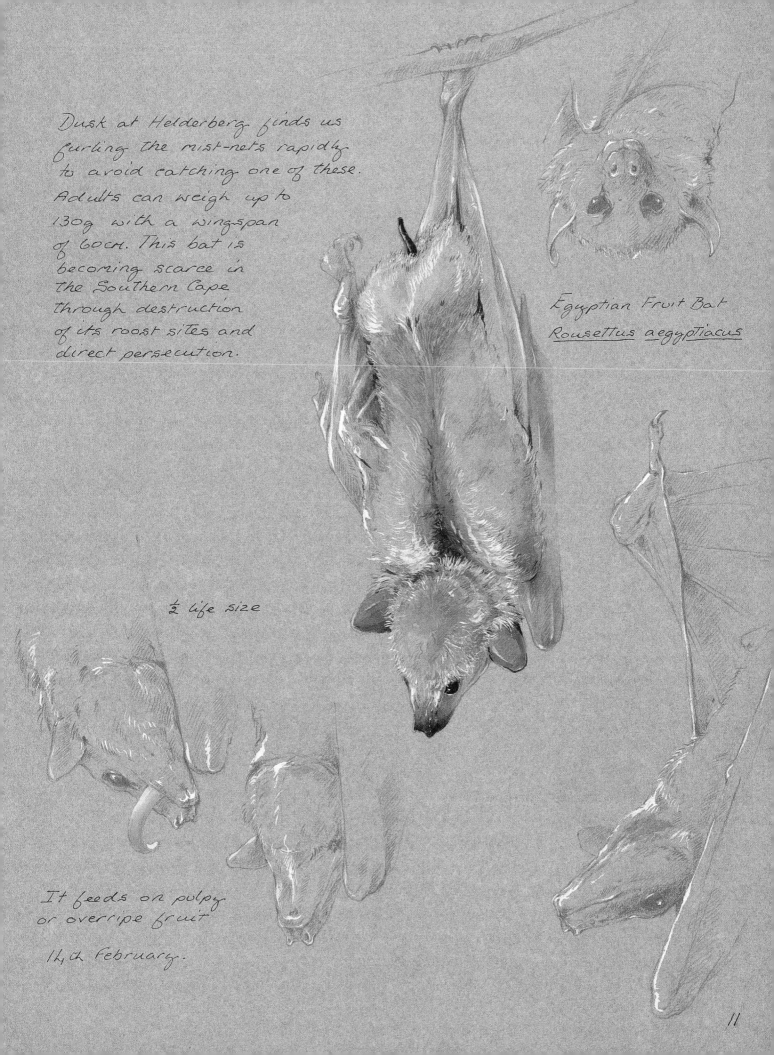

Dusk at Helderberg finds us
furling the mist-nets rapidly
to avoid catching one of these.
Adults can weigh up to
130g with a wingspan
of 60cm. This bat is
becoming scarce in
the Southern Cape
through destruction
of its roost sites and
direct persecution.

Egyptian Fruit Bat
Rousettus aegyptiacus

½ life size

It feeds on pulpy
or overripe fruit

14th February.

11

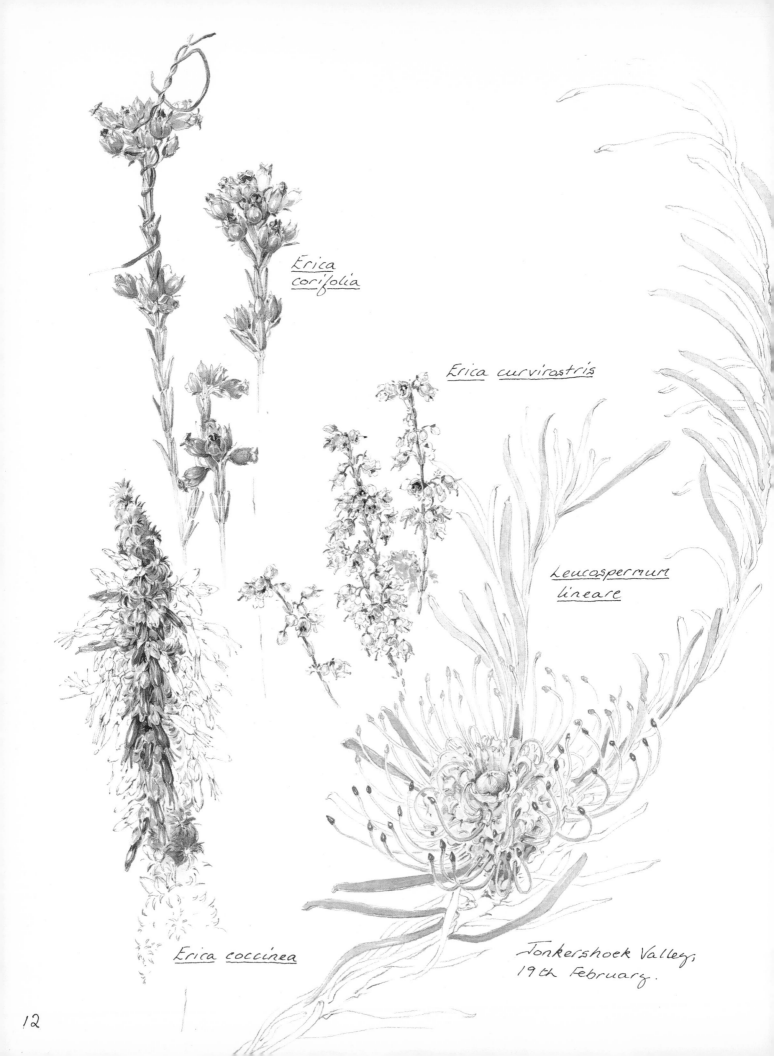

_Erica_
_corifolia_

_Erica curvirostris_

_Leucospermum_
_lineare_

_Erica coccinea_

Jonkershoek Valley,
19th February.

12

Anapalina nervosa (Karkarblom)

Found growing, sometimes as tall
as 75 cm, on mountain slopes in
the Jonkershoek valley. The
vivid scarlet of the flowers
contrasting with the sombre greens
and browns characteristic
of late-summer fynbos.

22nd February.

13.

The Rain Spider *Palystes* _natalius_. This old nest comprised fifty-eight *Brachylaena* _neriifolia_ leaves and one small Bracken frond

A brilliantly coloured silver vleispider (family Metidae) beside its tiny nest built among fern fronds.

7th February, Jonkershoek Valley.

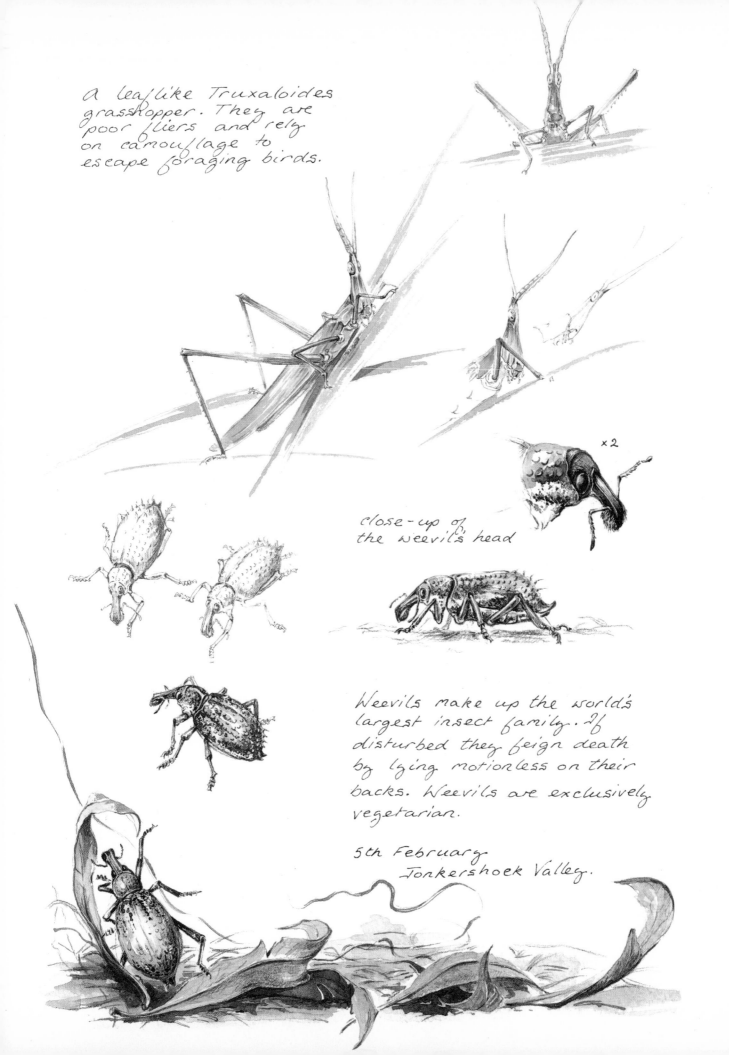

A leaflike Truxaloides grasshopper. They are poor fliers and rely on camouflage to escape foraging birds.

close-up of the weevil's head

×2

Weevils make up the world's largest insect family. If disturbed they feign death by lying motionless on their backs. Weevils are exclusively vegetarian.

5th February
Jonkershoek Valley.

15

A Cetoniid beetle
searching for nectar
in the flower-head.

unopened _P. nitida_
inflorescence (above)

_Protea nitida_
(Waboom) - the Wagon Tree
So called because the wood was once
used for making wagon-wheel rims.
Swartboskloof 25·2·87.

young leaf shoot

*Aspalathus crenata*

*Tritoniopsis dodii*
The only remaining plant
in flower, the others nearby
have already
set seed.

dried
seed-pods

*Pelargonium
tabulare*

A brightly-coloured
"Cotton-stainer"
beetle.

17

This blister beetle was busily feeding on the flowers of _Bobartia_ and _Anapalina_. I can understand why these insects are so unpopular with gardeners!

×2

Fire-break flowers

_Bobartia indica_ and a bright blue _Lobelia_ in the Tonkershoek Valley.

7.3.87

18

I was surprised to find this
beautiful <u>Roella ciliata</u> flowering
in the middle of a well-trodden
and dusty path at Swartboskloof.
The petals unfurled at nine o'clock
and closed again at two.

8.3.87

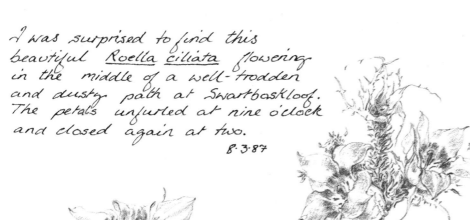

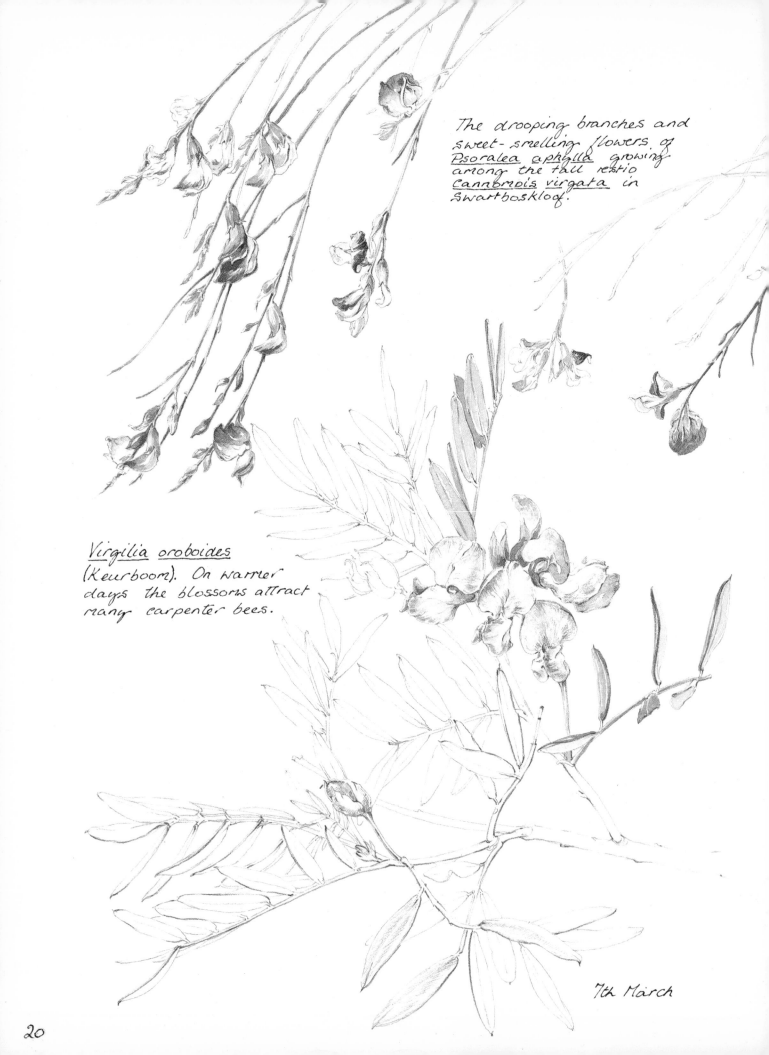

The drooping branches and
sweet-smelling flowers of
*Psoralea aphylla* growing
among the tall restio
*Cannomois virgata* in
Swartboskloof.

*Virgilia oroboides*
(Keurboom). On warmer
days the blossoms attract
many carpenter bees.

7th March

20

Langrivier, Jonkershoek Valley.
10th March

Decaying leaf-packs are an
important source of nutrients
in Mountain Fynbos streams.
Freshwater crabs _Potamon_
_perlatus_ glean food particles
from the leaves.

×2

Adult mayfly
Ephemeroptera

Mayfly nymphs

Lesser Doublecollared Sunbird
( Klein - rooiborssuikerbekkie )
*Nectarinia chalybea*
with colour rings.

Fine-mesh, almost invisible "mist nets" are set between poles to catch birds for ringing. Birds flying into the net form a deep "pocket" from which they are carefully extracted.

uniquely-numbered metal rings range in size from 1,8mm to 26,0mm.

Moult (feather replacement) is an important aspect of a bird's life. Here a Cape Weaver is moulting its flight feathers in sequence. Feathers are "scored" from 0 (old) to 5 (new, fully grown) according to age and stage of growth.

10
9
8
7
6
5
4
3 (missing)
2
1

Primaries

3 Tertials
6 secondaries

ring string

# A bird in the hand.

Once the bird is held firmly but gently in the "ringer's grip" a metal ring of the appropriate size is put on the leg with special pliers. Each bird is identified, aged, sexed, weighed, the lengths of tail, beak and folded wing measured and moult recorded.

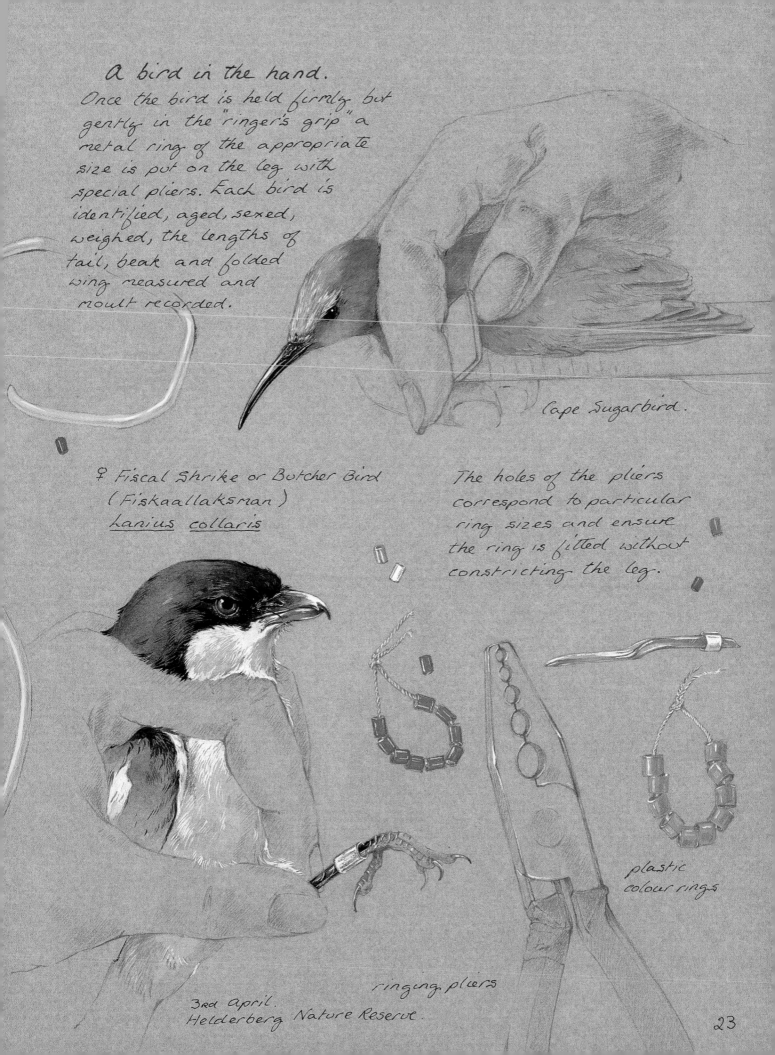

Cape Sugarbird.

♀ Fiscal Shrike or Butcher Bird
(Fiskaallaksman)
<u>Lanius collaris</u>

The holes of the pliers correspond to particular ring sizes and ensure the ring is fitted without constricting the leg.

plastic colour rings

ringing pliers

3rd April.
Helderberg Nature Reserve.

23

*Freylinia lanceolata*

Honeybell Bush
(Heuningklokkiesbos)

Jonkershoek Valley, 9th April.

*Gnidia oppositifolia*

24

The male of this species
has a beautiful blue-
green head.

Dozed quietly in
the warmth of my
hand. Released
after sketching.

Rock Agama _Agama atra_
When nervous or curious
agamas bob their heads rapidly
up and down — hence the
Afrikaans name 'koggelmannetjie'.

Cape of Good Hope Nature Reserve, 14th April

A Garden Spider
_Argiope_ sp. web-
spinning.

_Salvia africana_
Wild Sage.

Stellenbosch, 10th April.

26

Cape or Three-striped
Skink.
 Mabuya capensis

Knox's Ocellated
Sand Lizard
Meroles knoxii

Speckled or Orange-
legged Skink
Mabuya homalocephala

12th April.

27

Autumn flowers, Cape of Good Hope
Nature Reserve  13th - 16th April.

Despite the attentions of grazing
Bontebok, the vegetation in the
northern sector of the reserve is
showing some signs of recovery
a year after the wildfire.
These flowers were found in
a burnt area on the Atlantic
coast near Menskoppunt.

_Tritoniopsis dodii_

_Penaea mucronata_

_Roella ciliata_

_Erepsia gracilis_

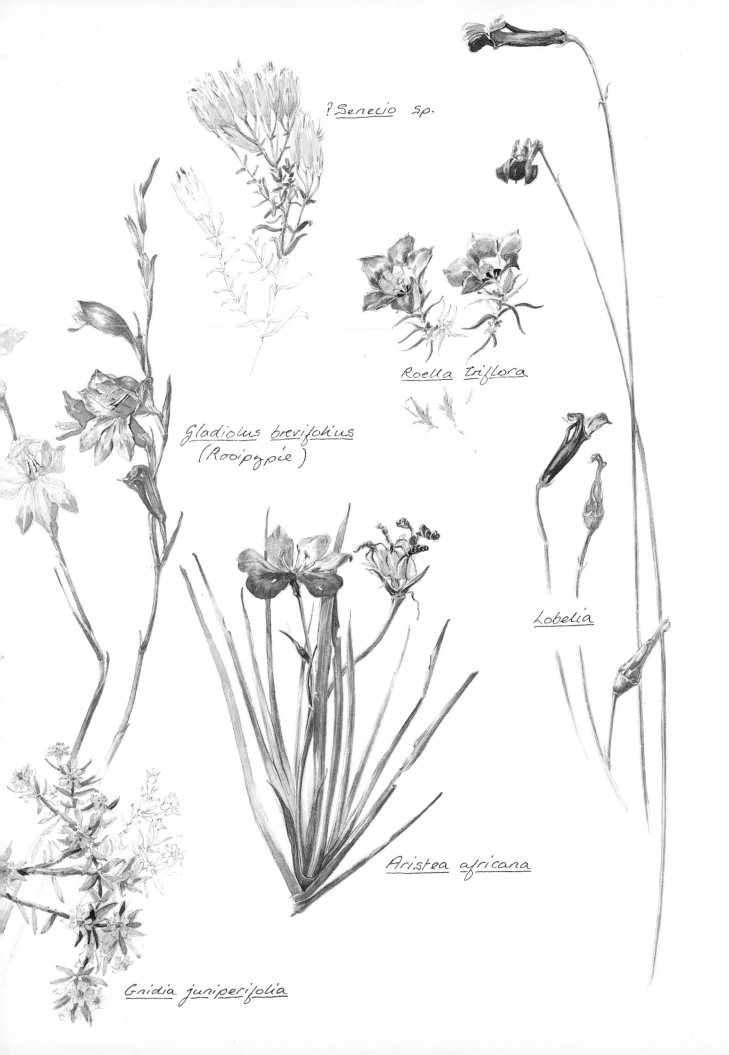

?Senecio sp.

Roella triflora

Gladiolus brevifolius
(Rooipypie)

Lobelia

Aristea africana

Gnidia juniperifolia

29

*Gladiolus brevifolius*

Cape of Good Hope
Nature Reserve.

15th April

30

Egyptian Goose, (Kolgans)
Alopochen aegyptiacus
Breeds in an old Hamerkop's
nest at Olifantsbos from where
the goslings are led down
to the sea. This species may
be found at almost every dam
and vlei.

Sirkelsvlei, Cape of Good Hope
Nature Reserve. 16th April.

31

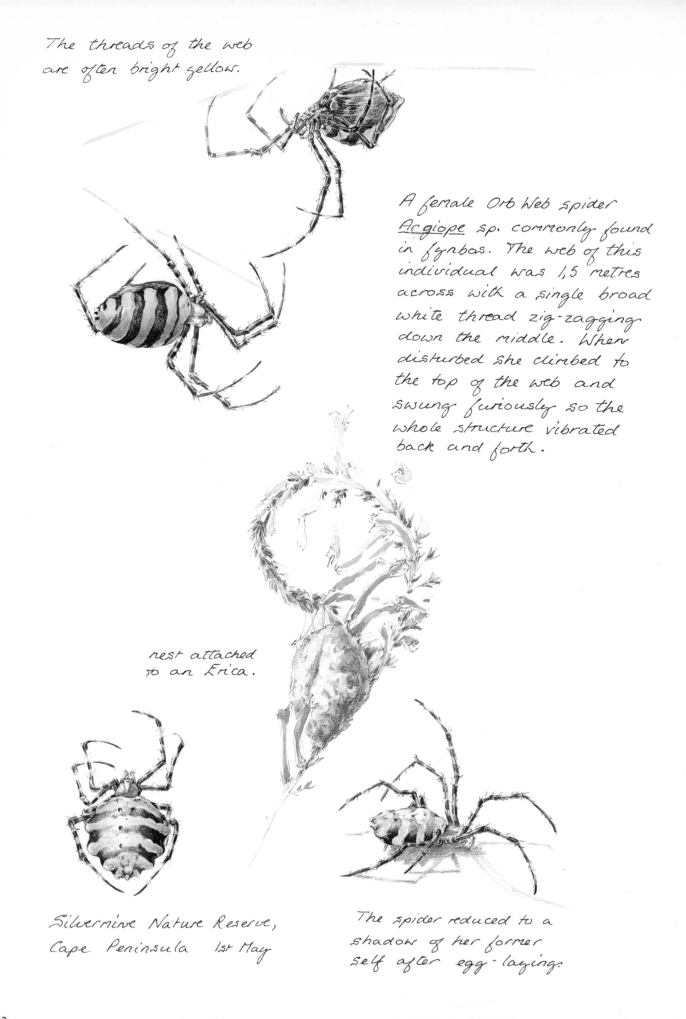

The threads of the web are often bright yellow.

A female Orb Web spider _Argiope_ sp. commonly found in fynbos. The web of this individual was 1,5 metres across with a single broad white thread zig-zagging down the middle. When disturbed she climbed to the top of the web and swung furiously so the whole structure vibrated back and forth.

nest attached to an Erica.

Silvermine Nature Reserve, Cape Peninsula 1st May

The spider reduced to a shadow of her former self after egg-laying.

Pine-tree emperor moth
*Imbrassia cytherea*

This impressive moth
hatched from a pupating
caterpillar collected at
Rocklands nine months ago

Wingspan almost 14cm.

A Convolvulus
hawk moth
*Agrius convolvuli*

A curious
unidentified
moth from
Stellenbosch

found dead at
Helderberg. The caterpillars
feed on Morning Glory.

3rd May

Dwarf Chameleon
*Microsaura pumila*
and her thirteen
offspring.
7th May, Stellenbosch.

34

Blushing Bride
<u>Serruria florida</u>

Protea Heights, Stellenbosch.
12th Mag.

35

This curious-looking
bird preys mainly on frogs
and their tadpoles. The
nest, an enormous chamber
of sticks and natural and artificial debris,
may be sited on a cliff-ledge or
in a tree.

braced against
the wind

Hamerkop *Scopus umbretta*
near Stellenbosch 16th May

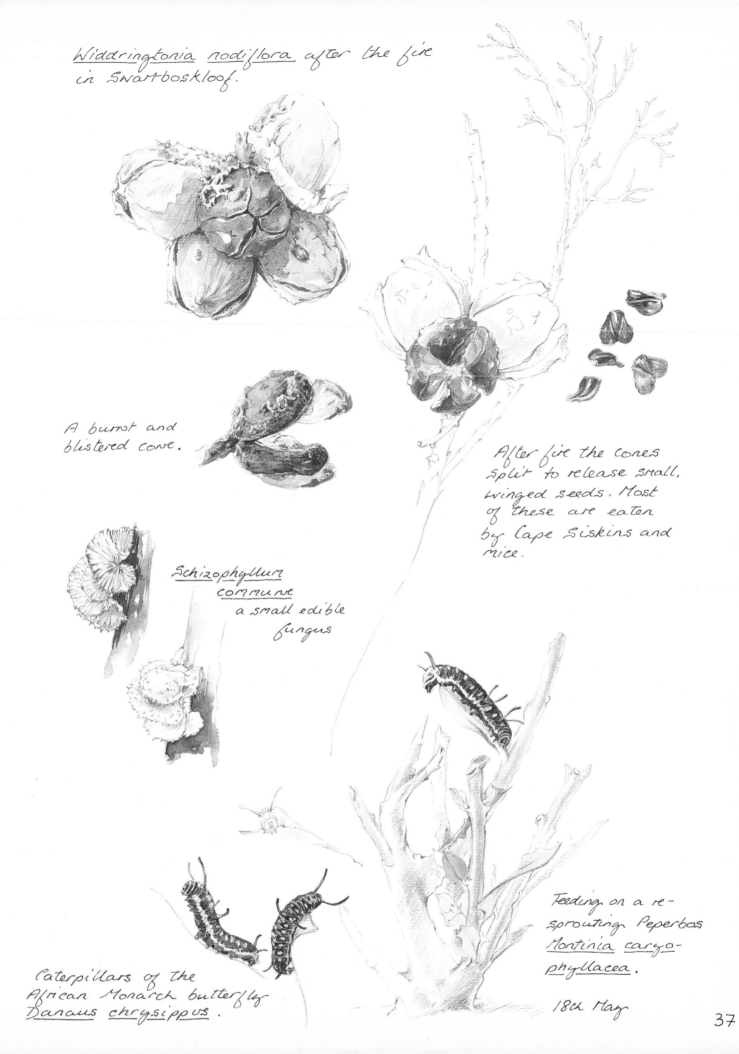

_Widdringtonia nodiflora_ after the fire in Swartboskloof.

A burnt and blistered cone.

After fire the cones split to release small, winged seeds. Most of these are eaten by Cape Siskins and mice.

_Schizophyllum commune_ a small edible fungus

Feeding on a re-sprouting Peperbos _Montinia caryo-phyllacea._

18th May

Caterpillars of the African Monarch butterfly _Danaus chrysippus._

37

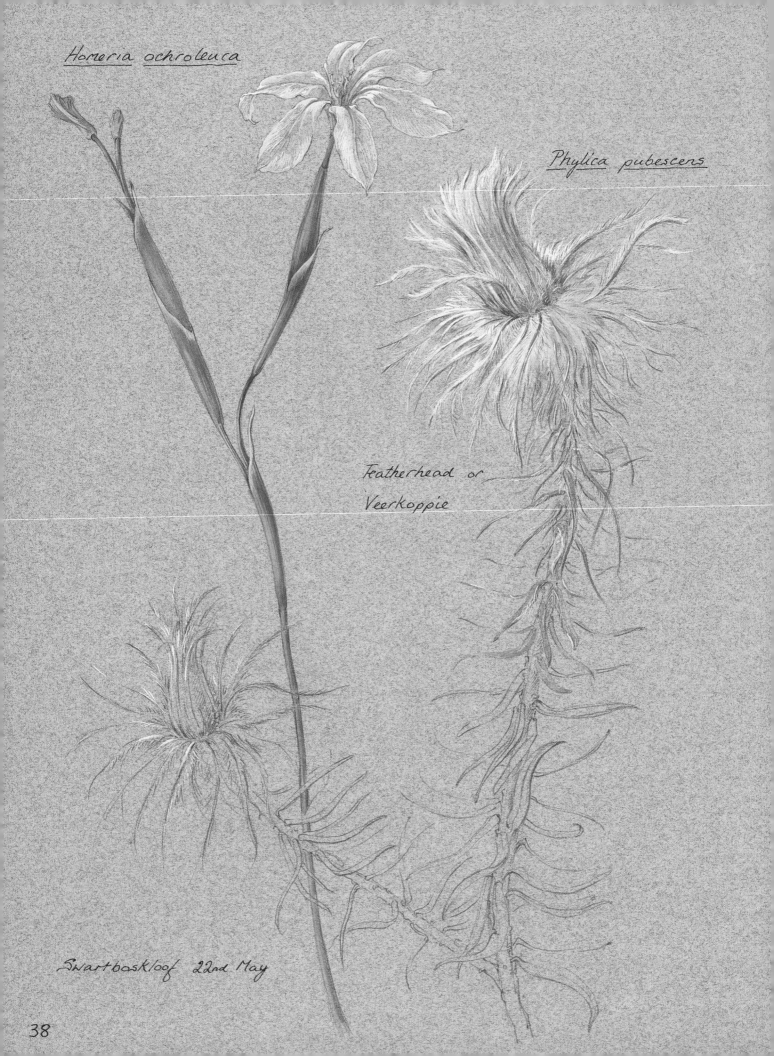

*Homeria ochroleuca*

*Phylica pubescens*

Featherhead or
Veerkoppie

Swartboskloof 22nd May

38

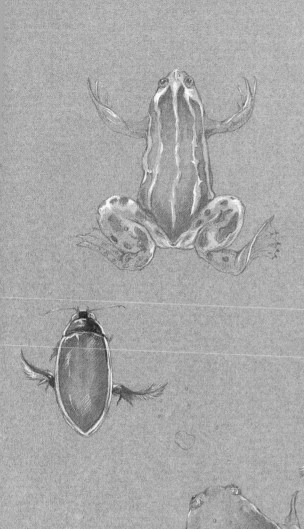

One of the rarest animals
in fynbos, the Cape Platanna
_Xenopus gilli_ is confined
to a scattering of black-
water lakelets in the
southwestern Cape.

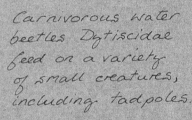

Carnivorous water
beetles Dytiscidae
feed on a variety
of small creatures,
including tadpoles.

The Common Platanna
_Xenopus laevis_ preys
on and breeds with
_X. gilli_, threatening it
with extinction.

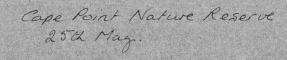

Cape Point Nature Reserve
25th May.

(Kapokbossie)

_Eriocephalus africanus_
Wild rosemary.

_Erica dichrus_
Pollinated by sunbirds; the
sticky flowers deter nectar-stealing
insects.

June 1987     Helderberg Nature Reserve

*Podalyria sericea*
Small, pink, pea-like
flowers.

*Leucadendron* sp.

Almost eighty
species occur in
fynbos.

*Leucadendron salignum.*

41

*Gladiolus maculatus*

Brown Afrikander.
Cape of Good Hope
Nature Reserve
9th June

42

Crowned Plover chick
found at Olifantsbos,
22nd June 1986 - three months
after the wildfire there.

Crowned Plover
Vanellus coronatus

The Kroonkiewiet is a familiar
bird of roadside verges and
playing fields in urban areas.
In fynbos this opportunist
species will take advantage
of a veld fire to feed on
invertebrate casualties and,
later, grasshoppers and other
insects attracted to the fresh
plant growth.

adult

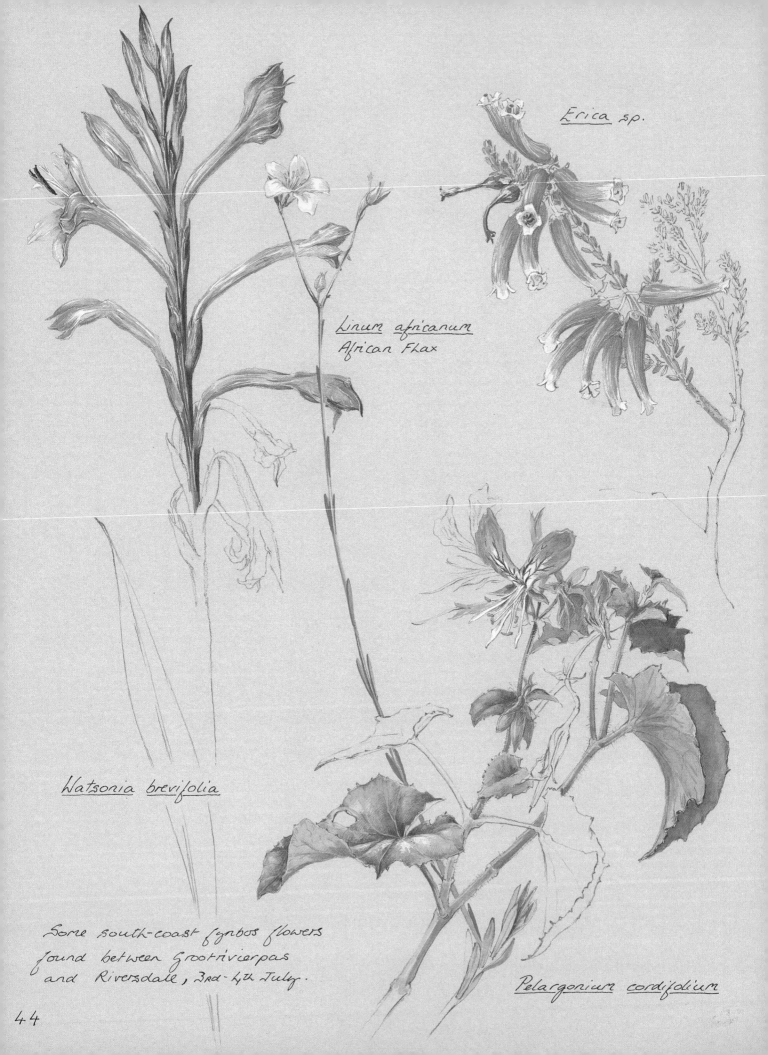

Erica sp.

Linum africanum
African Flax

Watsonia brevifolia

Pelargonium cordifolium

Some south-coast fynbos flowers
found between Grootrivierpas
and Riversdale, 3rd-4th July.

44

Knysna Dwarf Chameleon
Bradypodion damaranum
found in forested areas
and adjacent fynbos in
the southern Cape.

Knysna 5th July.

45

Grassbird
*Sphenoeacus afer*

A road casualty from
Ou Kaapse Weg, found
by David Allan on
6th July.

fiery-red
eye of
adult

breast
feather

back feather

downy semiplume
giving a "puffy" look to
the body

Wing    71mm
Tail    109mm
Tarsus  24,8mm
Culmen  12,6mm
Mass    27,5g

Found singly or in pairs in
dense vegetation, often in damp
areas. Characteristic short,
rounded wings and trailing
raggedy tail. A fine songster,
particularly vocal in the evening.

46

Grassbird.
*Sphenoeacus afer*

Creeping through marshy
vegetation at Mont Fleur,
Blaauklippen Valley.

10th July.

47

_Gladiolus carneus seed heads_

_Chasmanthe aethiopica_
(Suurkanolpypie)

Compositae sp.

short-horned grasshopper

48

Jonkershoek Valley July 14th

*Gladiolus carinatus*

The Sandpypie or Mauve Afrikaner
- a strong, almost sickly,
sweet fragrance.

*Boophane guttata* (Sambreelblom)

A dried inflorescence with ripening
seeds. 49

_Moraea tripetala_

flower from above

On the underside of every _Moraea_ flower a single brightly-coloured spider was found, lying in wait for any insect attracted to the nectar.

a straggling, white *Romulea*

*Jonkershoek Valley*   *15th July*

*Crassula capensis*
*Cape Snowdrop*

51

Corymbium villosum

White flowers also occur.
(Heuningbossie)

Swartboskloof, 20th July
A cold day with
snow on the mountains.

52

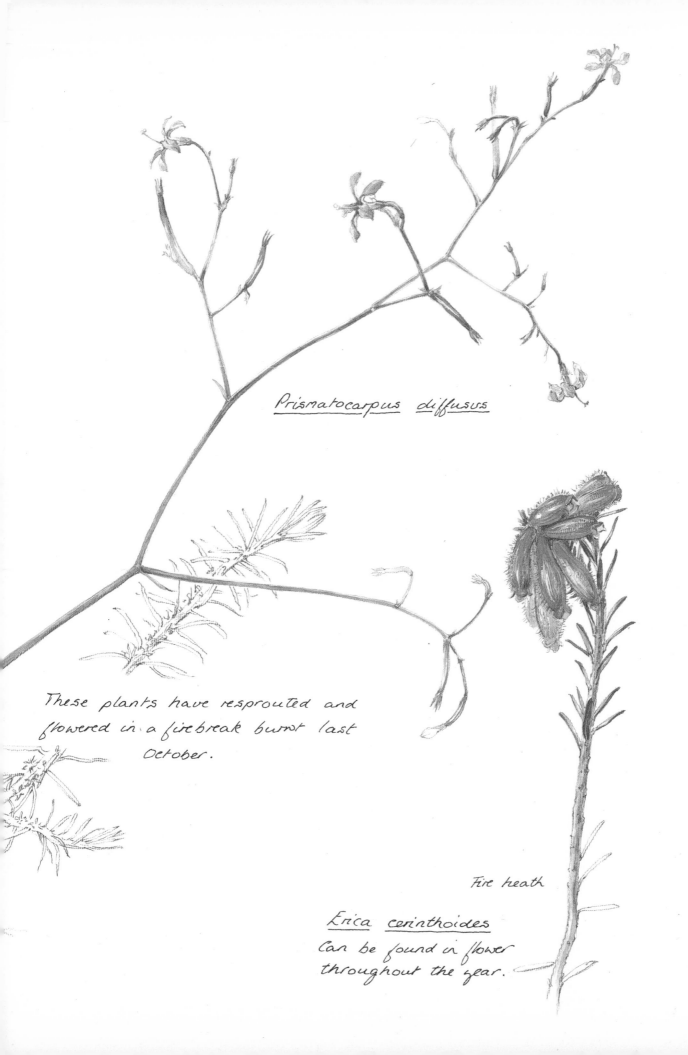

_Prismatocarpus diffusus_

These plants have resprouted and
flowered in a firebreak burnt last
October.

Fire heath

_Erica_ _cerinthoides_
Can be found in flower
throughout the year.

Protea lacticolor

Protea nana
(Skaamblom) Shy Flower

Leucospermum
cordifolium
Nodding Pincushion

54

*Protea burchellii*  (left and below)

*Protea eximia*
(Breëblaar-suikerbos)

Winter-flowering proteas
at Helderberg, 21st July.
Snow on the Hottentots
Holland mountains.

*Protea*
*scolymocephala*

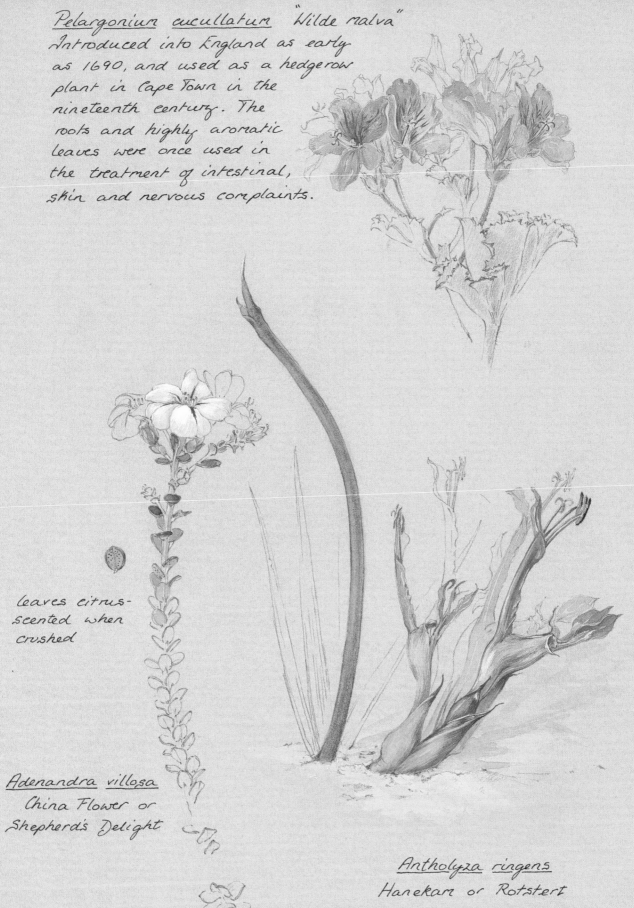

*Pelargonium cucullatum* "Wilde malva"
Introduced into England as early
as 1690, and used as a hedgerow
plant in Cape Town in the
nineteenth century. The
roots and highly aromatic
leaves were once used in
the treatment of intestinal,
skin and nervous complaints.

leaves citrus-
scented when
crushed

*Adenandra villosa*
China Flower or
Shepherd's Delight

*Antholyza ringens*
Hanekam or Rotstert

Three flowers from the Link Road,
Cape of Good Hope Nature Reserve.
The vegetation is recovering very slowly
after a wildfire swept through the
area in February, 1986.

1st August

56

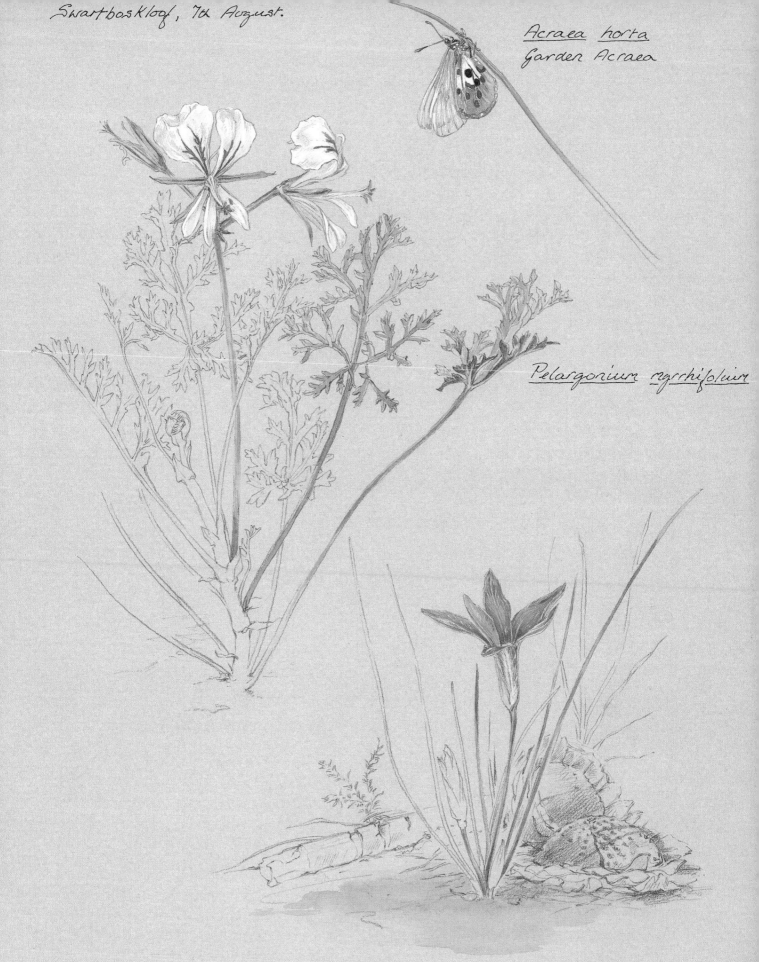

Swartboskloof, 7th August.

Acraea horta
Garden Acraea

Pelargonium myrrhifolium

Romulea rosea flowering among
charred protea heads.

Flowers and unripe
fruit of the Tortoise
Berry <u>Nylandtia</u> <u>spinosa</u>

<u>Zygophyllum</u> <u>procumbens</u>

<u>Zygophyllum</u> <u>flexuosum</u>
(Spekbossie) growing in
Dune Mixed Fynbos at
Olifantsbos Point.

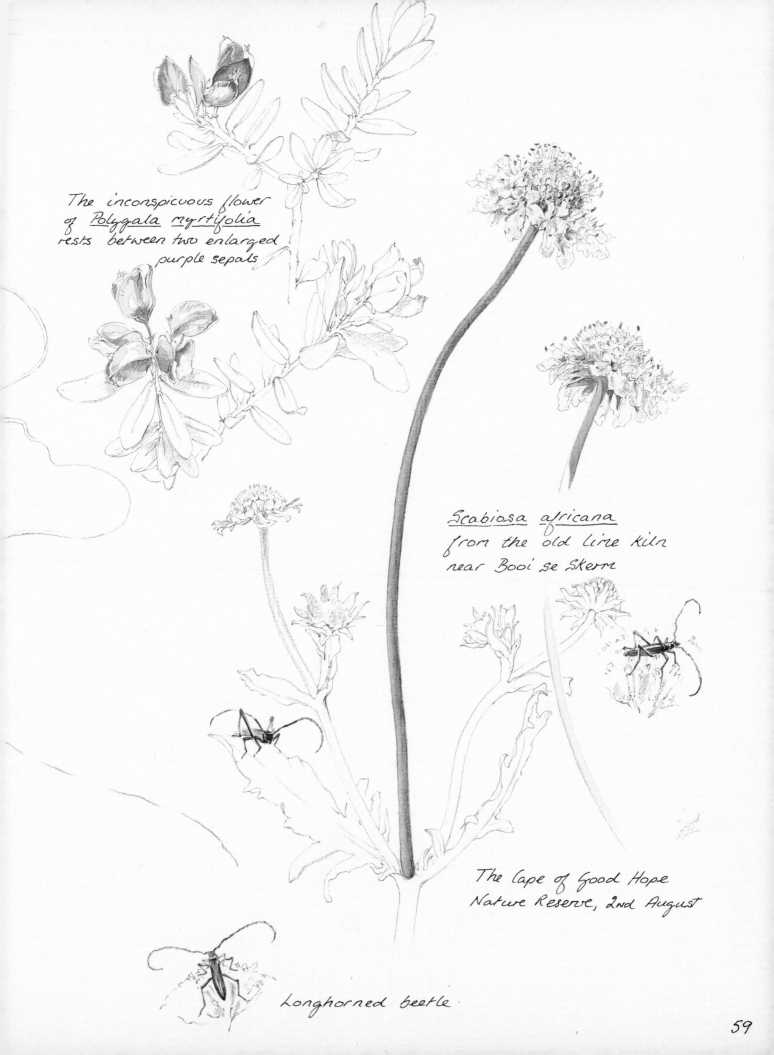

The inconspicuous flower of _Polygala myrtifolia_ rests between two enlarged purple sepals

_Scabiosa africana_ from the old lime kiln near Booi se Skerm

The Cape of Good Hope Nature Reserve, 2nd August

Longhorned beetle.

resting

A paper wasp's nest suspended
from a rocky overhang. Six
eggs and six larvae were found
in the cells. The grubs are
fed on chewed-up insects by
the female wasp.

This male is missing
a front leg

Malodorous twig-
wilter bugs _Anoplocnemis._
When provoked they secrete
a vile-smelling fluid to deter
would-be predators.

males with
inflated
hind femora

Hundreds of spikes of
_Lachenalia orchioides_ are
flowering along the
forest edge.

Unripe berries of the
fiendishly thorny
Protoasparagus

parasitic wasp

The higher slopes
of the kloof are dotted
with pale blue Aristeas

mating twig-wilters,
they spent three hours
in this position!

♂

♀

A tiny orange
fungus, swollen
after the rains.

Swartboskloof, 7th August

61

*Spiloxene capensis*
opening in the sun-
shine. The flowers are
very variable in size
(I found one over 10cm
across) and colour.

Jonkershoek, 21st August

Showy pink flowers of
Podalyria calyptrata,
the Waterkeurtjie.

Jonkershoek
Valley, 17 August

One of many wonderful
fynbos smells is that of
_Wurmbea spicata_ (below).
Its scent seems to be
particularly strong at
night - perhaps it is
pollinated by moths?

_Drosera trinervia_

The carnivorous sundew
supplements its nutrient
supply by trapping and
digesting insects on sticky
hairs covering the leaves.

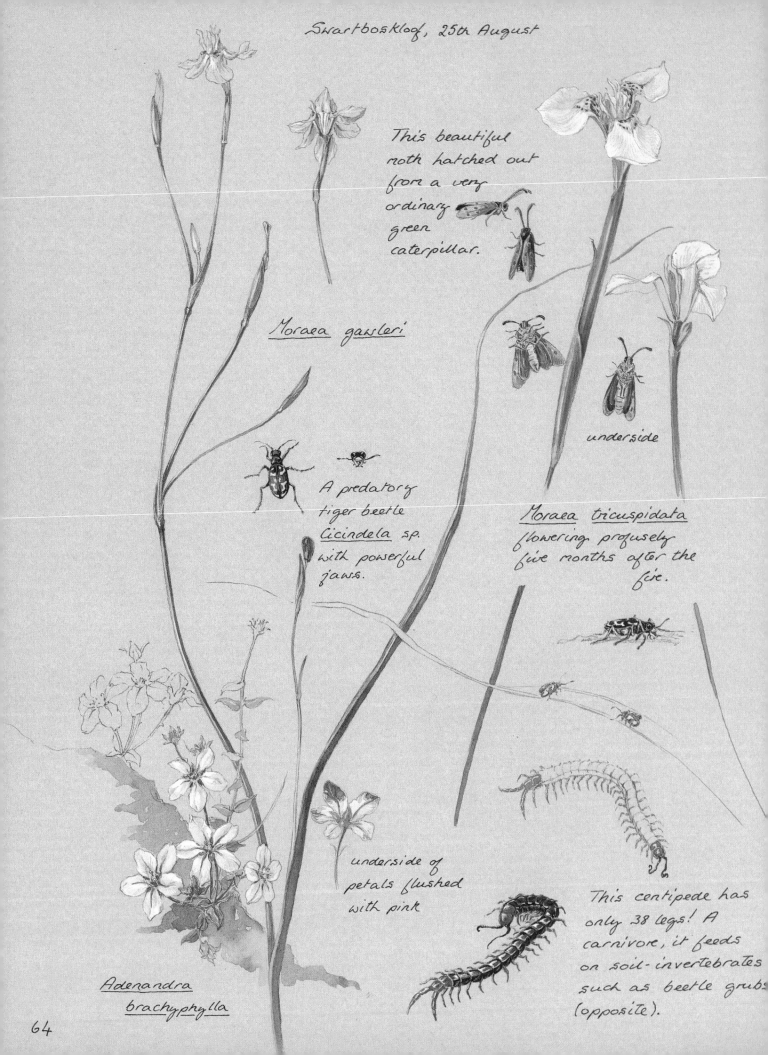

Swartboskloof, 25th August

This beautiful
moth hatched out
from a very
ordinary
green
caterpillar.

Moraea gawleri

A predatory
tiger beetle
Cicindela sp.
with powerful
jaws.

underside

Moraea tricuspidata
flowering profusely
five months after the
fire.

underside of
petals flushed
with pink

This centipede has
only 38 legs! A
carnivore, it feeds
on soil-invertebrates
such as beetle grubs
(opposite).

Adenandra
brachyphylla

64

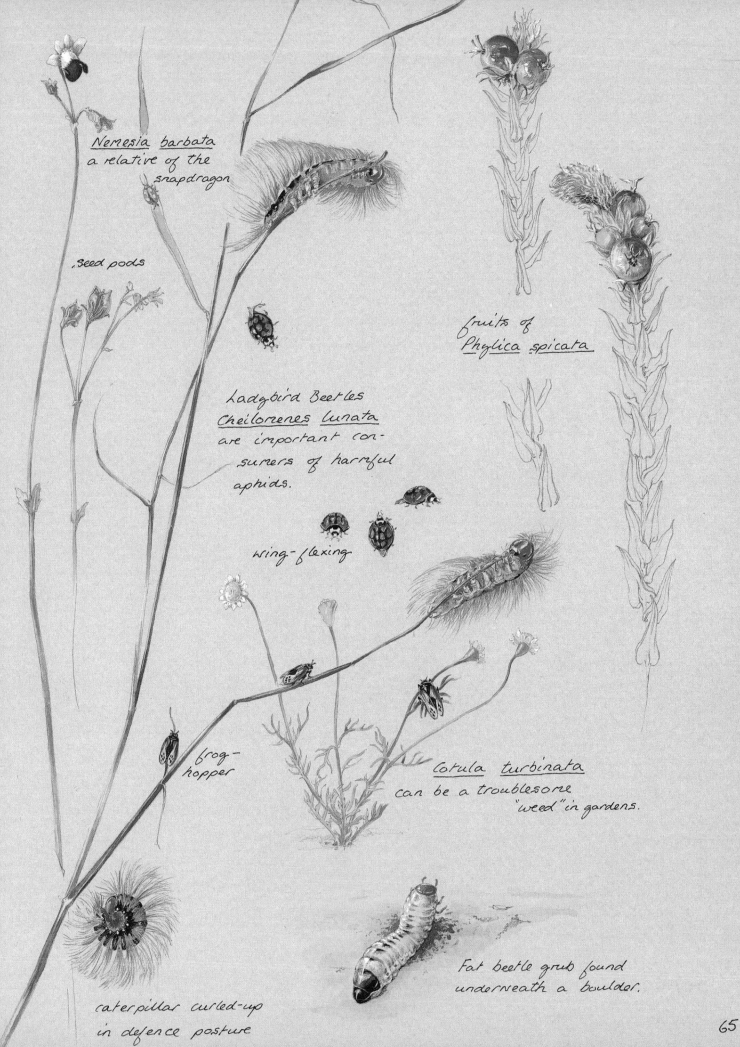

Nemesia barbata
a relative of the
snapdragon

seed pods

Ladybird Beetles
Cheilomenes lunata
are important con-
sumers of harmful
aphids.

wing-flexing

fruits of
Phylica spicata

frog-
hopper

Cotula turbinata
can be a troublesome
"weed" in gardens.

caterpillar curled-up
in defence posture

Fat beetle grub found
underneath a boulder.

Widespread in Mountain
Fynbos, the Rock Pigeon
has also readily adapted
to life in urban areas
where, to it, every building
is just another cliff.

Stellenbosch, 30th August.

Rock Pigeon (Kransduif)
Columba guinea

Stellenbosch, 31st August.

67

flowers close
by day

The flowers of
<u>Zaluzianskya dentata</u>
only open in the evenings.
Note the characteristically
long corolla tubes.

Praying mantid

<u>Crassula fascicularis</u>
The succulent and
fragrant Klipblom
is most sweet-
smelling at dusk.

68

calyx

petals

Restios tightly entwined
with _Cyphia volubilis_
(Aardboontjie)

_Hermannia_ sp.
with the petals
just protruding
from the hairy
inflated calyx.

This orchid,
_Pterygodium catholicum_
has a very strong soapy
smell. The bonnet-shaped,
green flowers give it the
Afrikaans name "Oumakappie."

I disturbed this
tiny pink
flower-spider.

A spike of _Cyphia
bulbosa_ (Bergbaroe),
very common in the
Kloof justnow.

_Pterygodium alatum_ is smaller
and the flowers less strongly-scented
than _P. catholicum_.

69

Spent a whole day mist-netting
in Swartboskloof. Almost 100 metres
of net caught a total of four
birds! This bright stocky
Bully Canary *Serinus sulphur-*
*atus* has a large and
powerful beak for husking
seeds and nipping ringers.
  (Dikbekkanarie)

A combination of colour rings,
unique to each bird, allows
individuals to be recognised
in the field.

9th September

Not a typical fynbos bird, the
Lesser Honeyguide *Indicator minor*
has apparently followed the spread
of its brood-host, the Pied Barbet
*Lybius leucomelas* into this habitat.

curious "port-
hole" nostrils

*Zygodactylic* —
two toes forward,
two toes back.

(Kleinheuningwyser)

Spotted Prinia

*Prinia maculosa*

A lively and excitable
little warbler, surprisingly
pugnacious in the hand!
(Karoolangstertjie)

Cape Batis

*Batis capensis* ♀ (Kaapse Bosbontrokkie)

Often ventures out of the
wooded Kloofs into adjacent
fynbos. Note distinctive black
mask.

An Albuca growing
among the thick,
fleshy leaves of
Sour Fig.

Olifantsbos, Cape
of Good Hope.
        10th September

Arum Lily Frog
*Hyperolius horstocki*

Rondevlei Bird Sanctuary
14th September.

Housed temporarily in a
clear tank, this Arum Lily
Frog _Hyperolius horstockii_
was quickly sketched and
then returned to its lily.

Suckers on all fingers
and toes enable this
little frog to cling firmly
to the sides of the glass,
displaying the bright
orange feet and hands.

A very attractive species which
is becoming increasingly rare through
the destruction of its wetland habitat.

When extended the legs are
surprisingly long and skinny.

The males have
a bright yellow vocal
sac under the chin.

Rondevlei Bird Sanctuary
14th September.

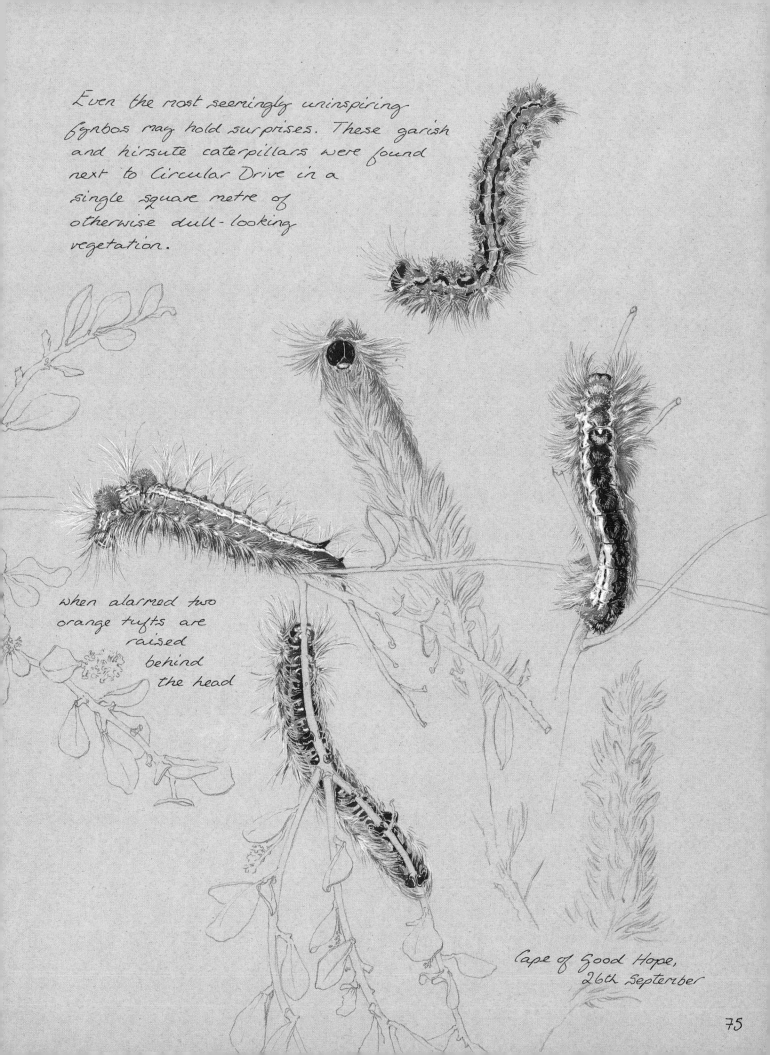

Even the most seemingly uninspiring
fynbos may hold surprises. These garish
and hirsute caterpillars were found
next to Circular Drive in a
single square metre of
otherwise dull-looking
vegetation.

when alarmed two
orange tufts are
raised
behind
the head

Cape of Good Hope,
26th September

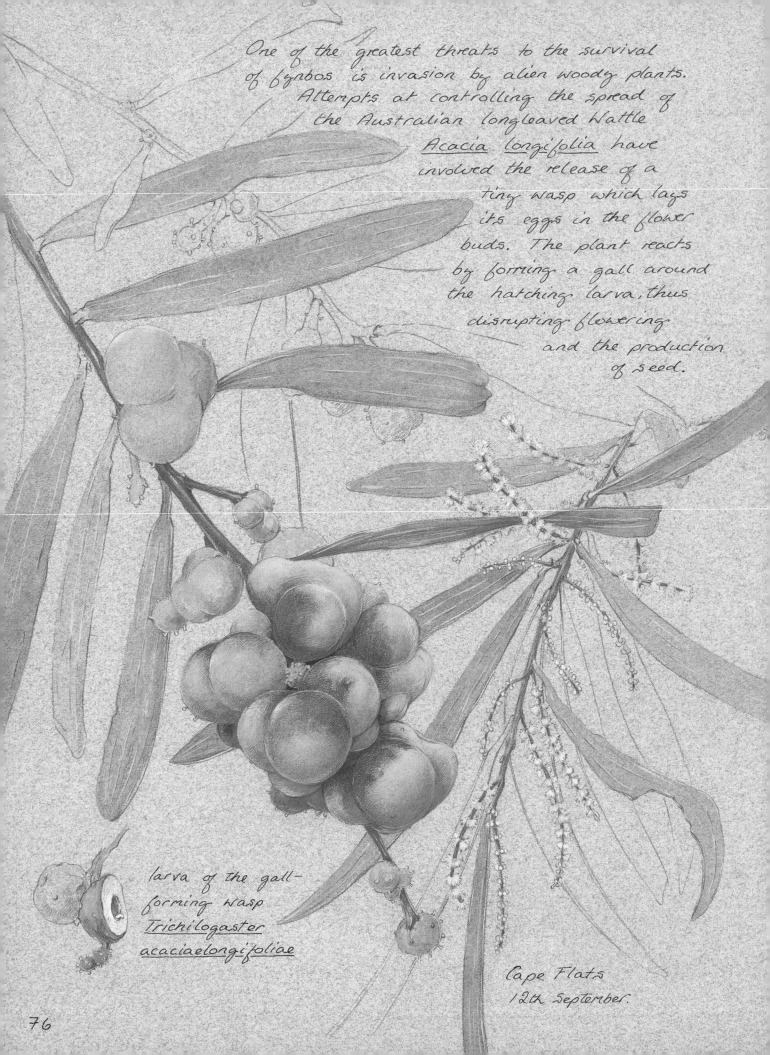

One of the greatest threats to the survival
of fynbos is invasion by alien woody plants.
Attempts at controlling the spread of
the Australian longleaved Wattle
_Acacia longifolia_ have
involved the release of a
tiny wasp which lays
its eggs in the flower
buds. The plant reacts
by forming a gall around
the hatching larva, thus
disrupting flowering
and the production
of seed.

larva of the gall-
forming wasp
_Trichilogaster_
_acaciaelongifoliae_

Cape Flats
12th September.

76

Cape Sugarbird nestling
about ten-days old. Most
of the Helderberg broods
had fledged by the end
of August, so this was
quite a late nest. Nestlings
solicit food from their parents
by calling frantically and
displaying their brightly-
coloured gapes.

18th September.

77

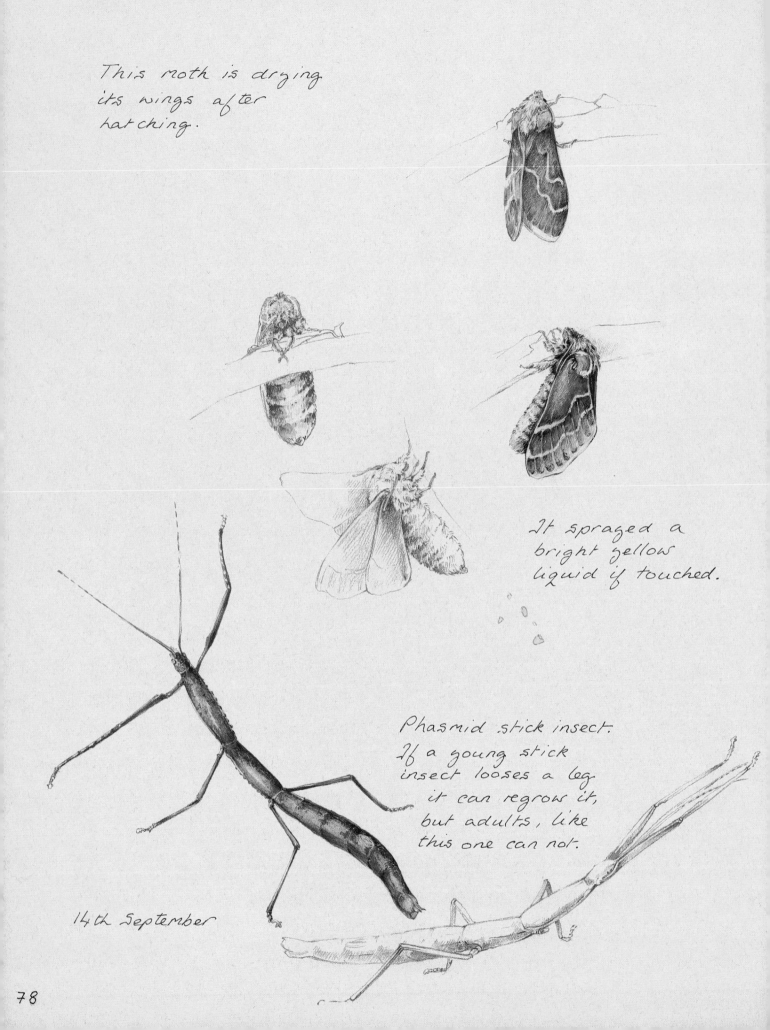

This moth is drying
its wings after
hatching.

It sprayed a
bright yellow
liquid if touched.

Phasmid stick insect.
If a young stick
insect looses a leg
it can regrow it,
but adults, like
this one can not.

14th September

78

Two goslings, over a
month old but still
very awkward on
their feet.
28th September.

The famous Spurwinged
Goose _Plectropterus_
_gambensis_ (Wildemakou),
which has nested on
the slopes of the Helder-
berg and led her broods
to the nature reserve
pond every spring for
almost fifteen years.

Nestling birds should be ringed only when the foot is big enough to prevent the closed ring falling off. This Fiscal Shrike chick (ring no. 480078) is the ideal age.

Helderberg 8th October.

One week later the chick is well-feathered and very lively. Such birds should never be disturbed as they are likely to "explode" from the nest before they can fly properly.

The nest is built mainly of _Eriocephalus_ (Kapokbossie) twigs and lined with tissue paper and sweet wrappers!

The curious Spinnekopblon
_Ferraria Crispa_ growing
at Cape Point, the south-
west tip of Africa. The
flowers are heavily
vanilla-scented and
last only one day.

Brilliant red _Anomolesia_
_cunnonia_ in coastal
dunes near Olifantsbos.

Cape of Good Hope Nature Reserve,
26th September.

81

Satyrium orchids from
the Gydo Pass (pink) and
burnt veld near Franschhoek
(white). The butterfly is a
Brown-veined white
Beleonois aurota

18th October.

82

An odd mammal, the dassie
has a digestive system
like a bird, teeth like a
rhinoceros and its
closest relatives are
the dugong and the
elephant!

Table Mountain
12th October.

A 110mm long Cape Rana
_Rana fuscigula_ tadpole
found near Ceres. Complete
metamorphosis can take
as long as three years.

The front right
leg has still to
emerge.

More respectably-sized
_Rana sp._ tadpoles from
Jonkershoek.

18th October

84

Also known as the
Parrot-beaked
Tortoise. The red nose
and bleary eyes give
it a distinctly
boozy look!

Southern or Areolate Padloper
(Holskubpadloper) <u>Homopus</u>
<u>areolatus</u> from the Therons-
bergpas, between Ceres
and Karoopoort.

19th October.

Twenty-two weeks after the
fire the first <u>Watsonia
pyramidata</u> (Suurkanol)
flowered at Swartboskloof. The blue
<u>Aristeas</u> followed soon after and
by mid-October the lower slopes were a
magnificent patchwork of pink and blue.

24th October

87

A recently fledged Cape Sugarbird,
still with its bright-yellow gape
flange. Feeding on pincushions.
Betty's Bay 30th October.

_Wachendorfia_
_paniculata_

(Rooikanol,
Spinnekopblom)
Swartboskloof,
            29th October.

day-flying moth

A small (20 cm)
Watsonia from
Betty's Bay,
            31st October.

The curious but endearing
Cape Mountain Rain Frog
_Breviceps montanus_ (Blaasop)
lives on stony, scrub-covered
hillsides and never
enters water. The tadpoles
develop in egg-capsules
in underground
chambers.

Inflates itself like
a puff-ball when
agitated.

A Cape Rana _Rana fuscigula_
swallows a grasshopper and then
looks decidedly dyspeptic
(below). This species is
found throughout fynbos
but is restricted to
permanent water bodies.

3rd November

90

An unidentified member
of the Iridaceae from
Betty's Bay.
7th November

Edmondia
sesamoides
(Strooiblommetjie)
a yellow
Everlasting.

91

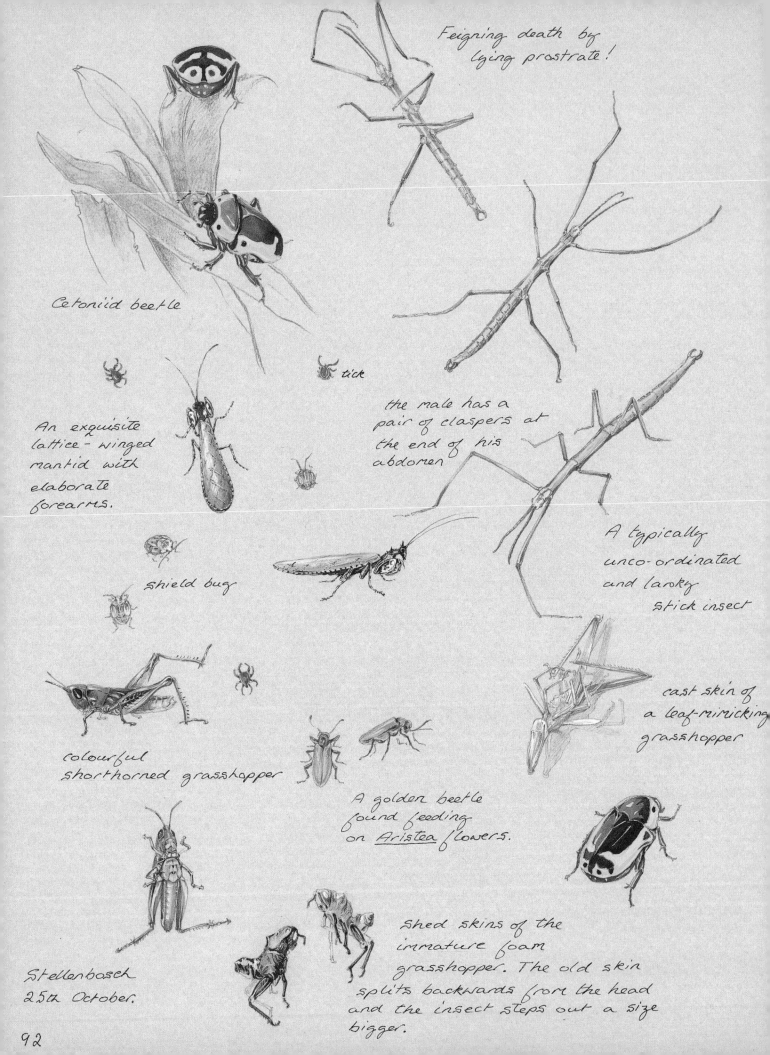

Feigning death by
lying prostrate!

Cetoniid beetle

tick

An exquisite
lattice-winged
mantid with
elaborate
forearms.

shield bug

the male has a
pair of claspers at
the end of his
abdomen

A typically
unco-ordinated
and laroky
stick insect

colourful
shorthorned grasshopper

A golden beetle
found feeding
on _Aristea flowers._

cast skin of
a leaf-mimicking
grasshopper

Shed skins of the
immature foam
grasshopper. The old skin
splits backwards from the head
and the insect steps out a size
bigger.

Stellenbosch
25th October.

92

Feeds on a variety
of fynbos fruits
and seeds, part-
icularly protea
such as _P. neriifolia_
illustrated here.

Swartboskloof
12th November.

Protea Canary
_Serinus leucopterus_

Rather an unspectacular
bird, but a real fynbos
speciality, found no-
where else. An impressive
mimic, one bird in Swart-
boskloof included in its
repertoire the calls of
Victorin's Warbler, Spotted Prinia,
Barthroated Apalis, Cape White-
eye (song and alarm call), Cape
Sugarbird, Orangebreasted Sunbird,
Cape Robin and a canary-like
twitter which may have been its own
song!

93

A selection of birds from
Olifantsbos and Helderberg
ringing days.
November

Pied Barbet
*Lybius leucomelas*
(Bonthoutkapper)
×2

Bokmakierie
*Telophorus
zeylonus*

Cape Bulbul
*Pycnonotus
capensis*
(Kaapse tiptol)

Wild Peach
*Kiggelaria africana*
(Wildeperske)
Eaten by bulbuls.

Yellowrumped Widow
*Euplectes capensis*
(Kaapse Flap)
Male in transitional plumage.

Cape Robin
*Cossypha caffra*
(Gewone Janfrederik)

Orangebreasted
Sunbird *Nectarinia*
*violacea* ♂
(Oranjeborssuikerbekkie)

Cape Weaver ♂
*Ploceus capensis*
(Kaapse Wewer)

Speckled Mousebird
*Colius striatus*
(Gevlekte Muisvoël)

Common Waxbill ♂
*Estrilda astrild*
(Rooibeksysie)

Hexaglottis virgata
(Volstruisuintjie)

Lycus beetle

Disa atricapilla

Ixia polystachya

Jonkershoek Valley,
24th November.

Ixia polystachya
(Koringblommetjie)

96

*Tritoniopsis*
*parviflora*

*Albuca*
*cooperi*

*Moræa ramosissima*
(*Geeltulp*)

*Cyanella hyacinthoides*
(*Raaptol*) *Ladg's Hand.*

97

Bat-eared Fox. (Bakoorvos)
*Otocyon megalotis*

near Langebaan, 27th November.

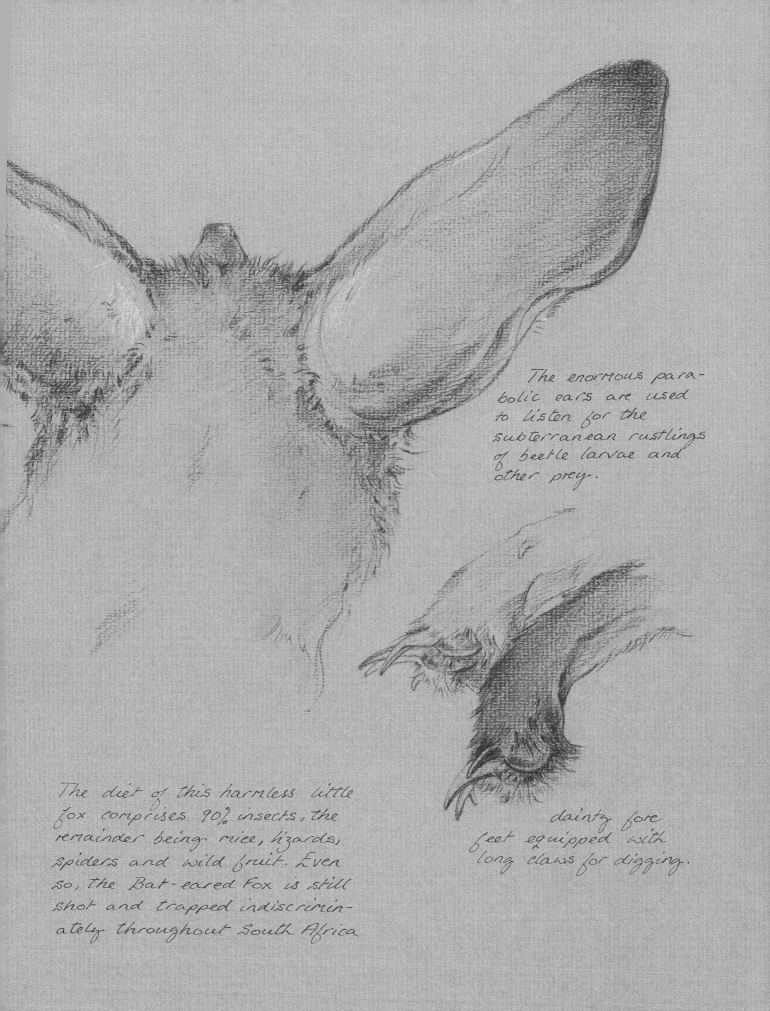

The enormous para-
bolic ears are used
to listen for the
subterranean rustlings
of beetle larvae and
other prey.

The diet of this harmless little
fox comprises 90% insects, the
remainder being mice, lizards,
spiders and wild fruit. Even
so, the Bat-eared Fox is still
shot and trapped indiscriminately throughout South Africa

dainty fore
feet equipped with
long claws for digging.

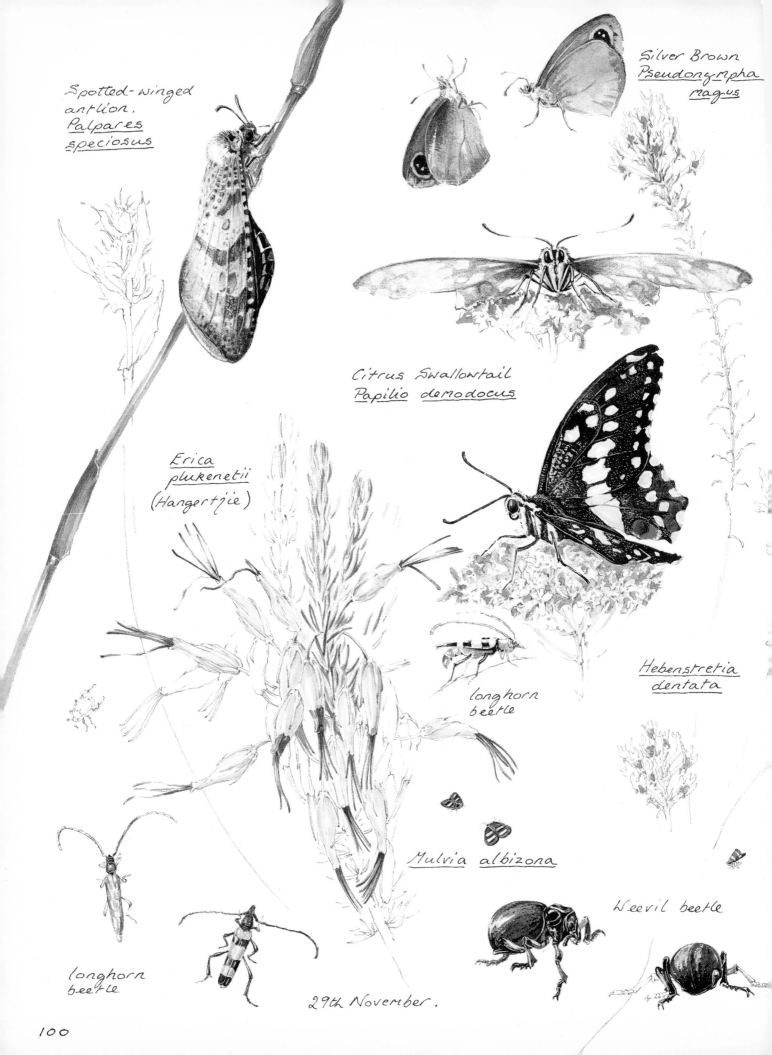

Spotted-winged
antlion.
*Palpares
speciosus*

Silver Brown
*Pseudonympha
magus*

Citrus Swallowtail
*Papilio demodocus*

*Erica
plukenetii*
(Hangertjie)

longhorn
beetle

*Hebenstretia
dentata*

*Mulvia albizona*

Weevil beetle

longhorn
beetle

29th November.

100

Vespid wasp
(paper wasp)

Blister beetle

Tabanid fly

lycaenid
butterfly

tortoise
beetle

Sphaecid thread-
waist wasp

♀
ovipositor

Disa patens

Cicada nymphs moult
while clinging to grass
stems.

Cicada adult.

monkey beetles

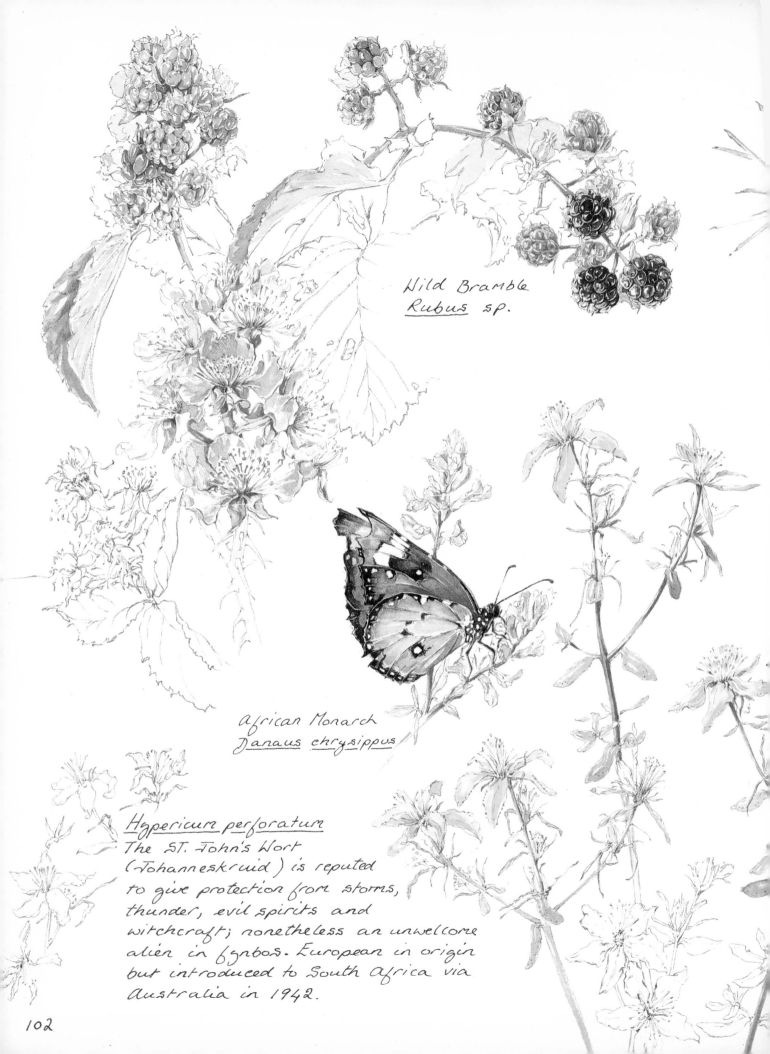

Wild Bramble
*Rubus* sp.

African Monarch
*Danaus chrysippus*

*Hypericum perforatum*
The St. John's Wort
(Johanneskruid) is reputed
to give protection from storms,
thunder, evil spirits and
witchcraft; nonetheless an unwelcome
alien in fynbos. European in origin
but introduced to South Africa via
Australia in 1942.

_Hakea sericea_ Silky (Syerige) Hakea
This notorious Australian poses an
enormous threat to fynbos and is
being countered by mechanical
and biological control measures.

winged hakea
seeds

Disturbed ground by the
Kleinplaas Dam, Jonkershoek,
provides suitable
habitat for lucerne
_Medicago sativa_, an
introduced fodder-plant.
The African Clouded
Yellow or Lucerne Butter-
fly _Colias electo_ uses
it extensively for feeding
and breeding.

6th December.

103

*Agapanthus africanus* (Bloulelie)
is one of many fynbos
flowers now widely cultivated.
It even flourishes in gardens
in the west of Scotland!

Swartboskloof 14th December.

104

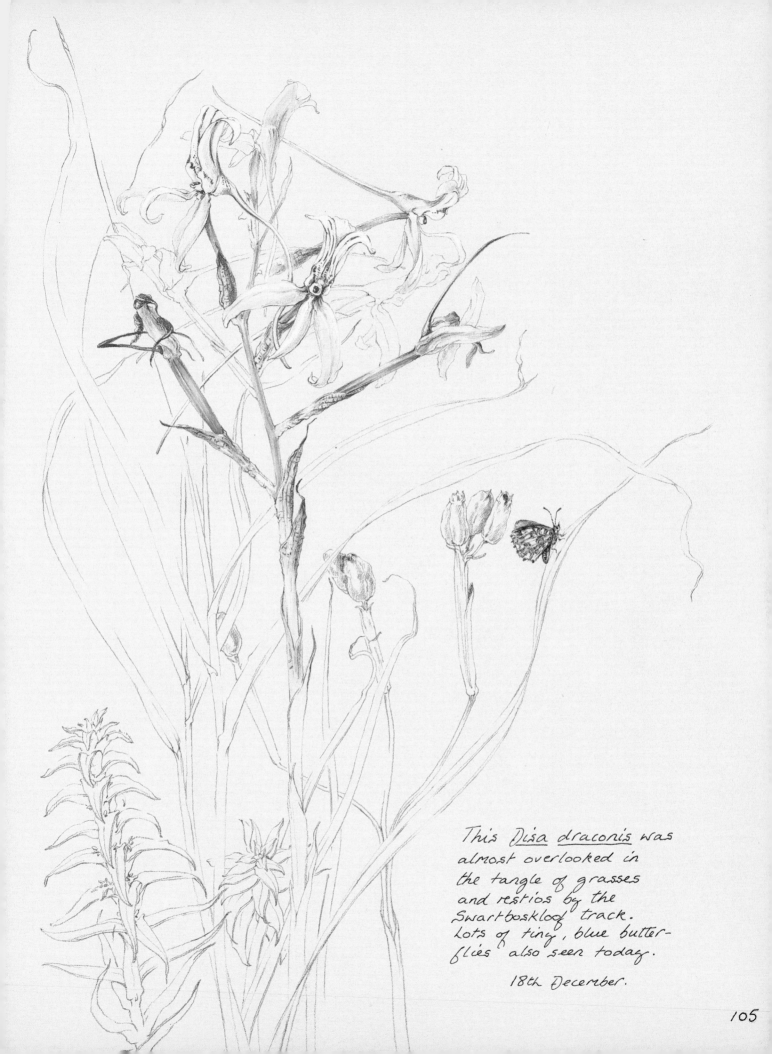

This *Disa draconis* was
almost overlooked in
the tangle of grasses
and restios by the
Swartboskloof track.
Lots of tiny, blue butterflies also seen today.

18th December.

_Witsenia maura_ (Bokmakieriestert)
Named in honour of Nicolaas
Witsen, a director of the Dutch
East India Co., and patron of
botanical research in early
18th century South Africa.
This rare and curious species,
  the "Bokmakierie's Tail", may
    now be found only at
      one or two sites on
        the Cape Peninsula.

20th December.

106

Egg shell and carapace
of an Angulate Tortoise
_Chersine angulata_
(Ploegskaarskilpad).

Cape Terrapin _Pelomedusa
subrufa_ (Gewone Waterskilpad)
shell.

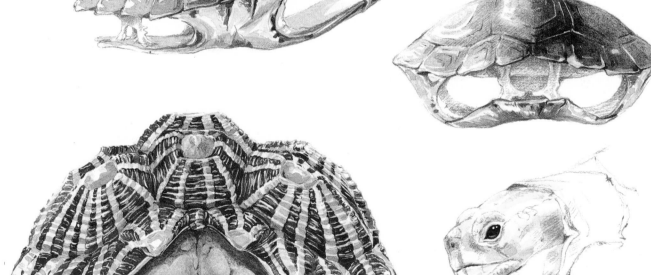

One of the world's
rarest tortoises - the
Geometric _Psammobates
geometricus_ (Suurpootjieskilpad).
Destruction of west-coast Renosterveld has
almost completely extinguished this beautiful
species.

26th December

We reunited this Egyptian
gosling with its eleven
siblings after finding it
on the wrong side of the
fence round the Helderberg
pond.

23rd December.

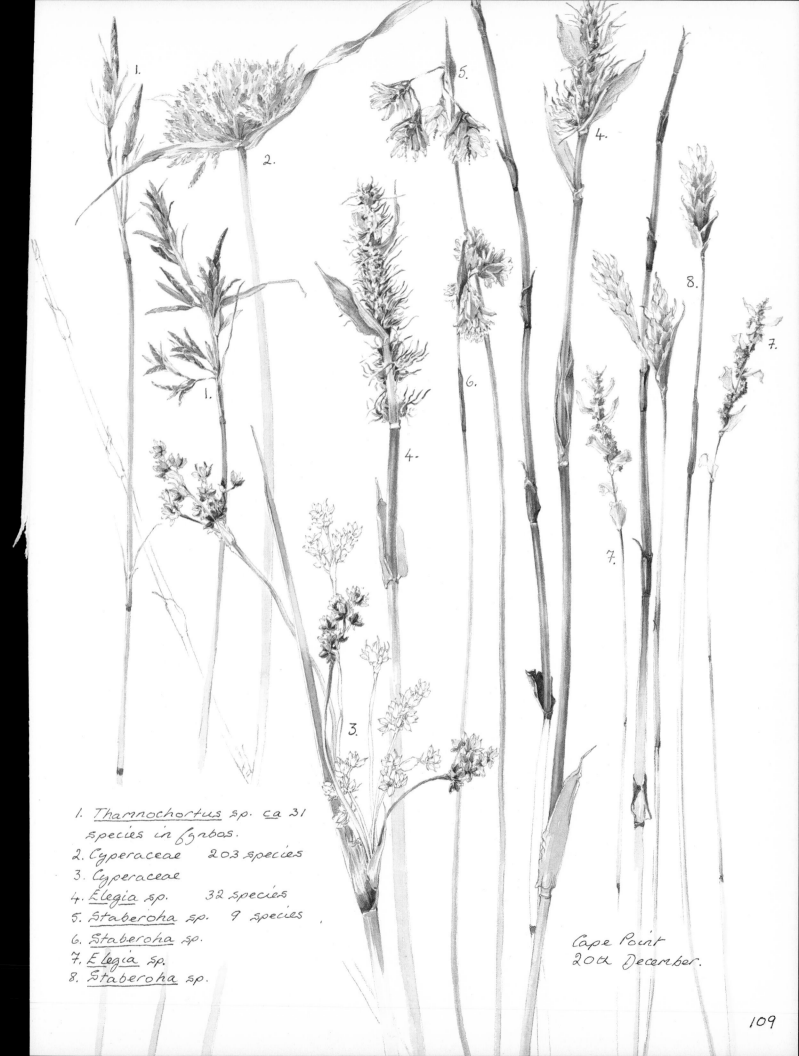

1. _Thamnochortus_ sp. ca 31
   species in fynbos.
2. Cyperaceae    203 species
3. Cyperaceae
4. _Elegia_ sp.      32 species
5. _Staberoha_ sp.   9 species
6. _Staberoha_ sp.
7. _Elegia_ sp.
8. _Staberoha_ sp.

Cape Point
20th December.

109

Large-spotted Genet
Genetta tigrina
(Rooikolmuskejaatkat)

Klaasjaggersberg, 27th December.

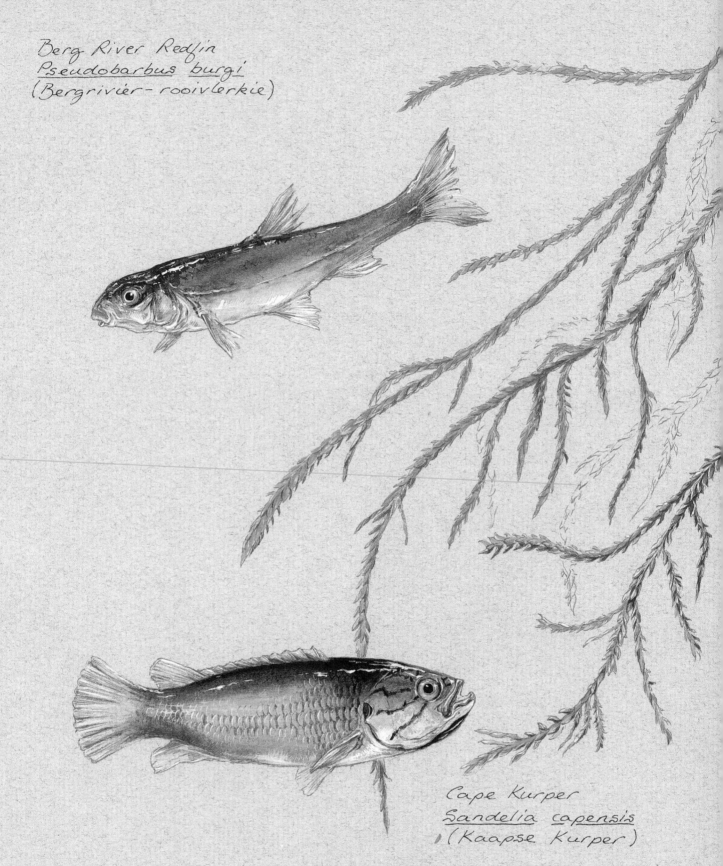

Berg River Redfin
*Pseudobarbus burgi*
(Bergrivier-rooivlerkie)

Cape Kurper
*Sandelia capensis*
(Kaapse Kurper)

29th December.

111

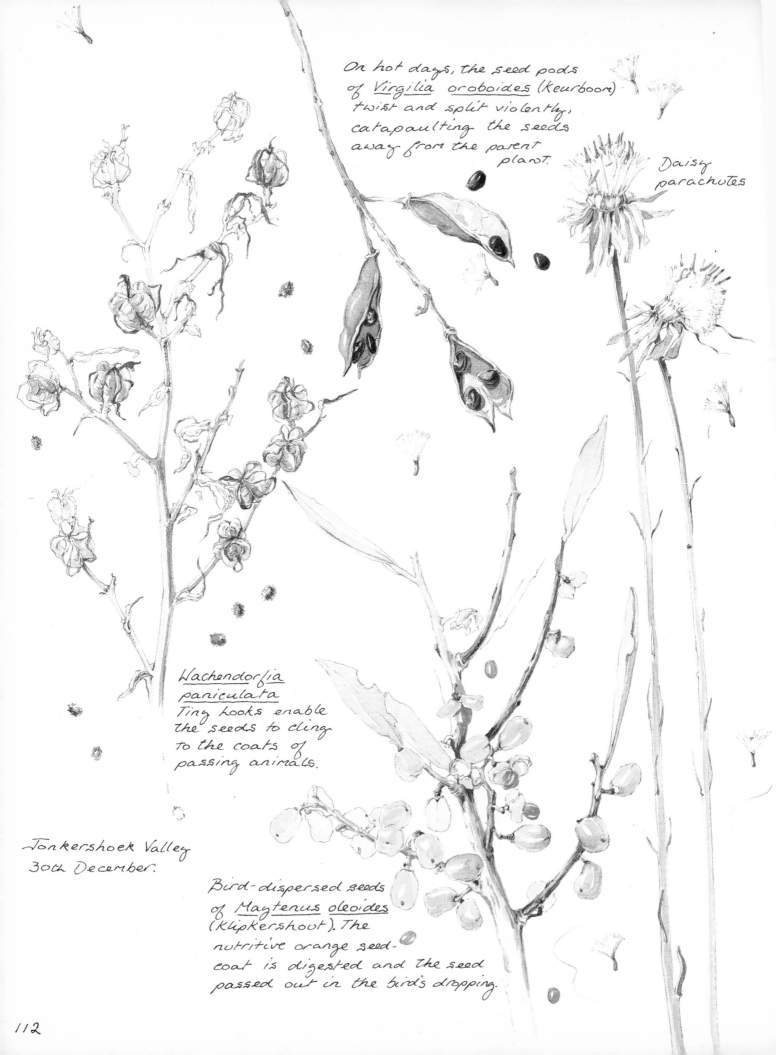

On hot days, the seed pods
of _Virgilia_ _oroboides_ (Keurboom)
twist and split violently,
catapaulting the seeds
away from the parent
plant.

Daisy
parachutes

_Wachendorfia_
_paniculata_
Tiny hooks enable
the seeds to cling
to the coats of
passing animals.

Jonkershoek Valley
30th December.

Bird-dispersed seeds
of _Maytenus_ _oleoides_
(Klipkershout). The
nutritive orange seed-
coat is digested and the seed
passed out in the bird's dropping.

112

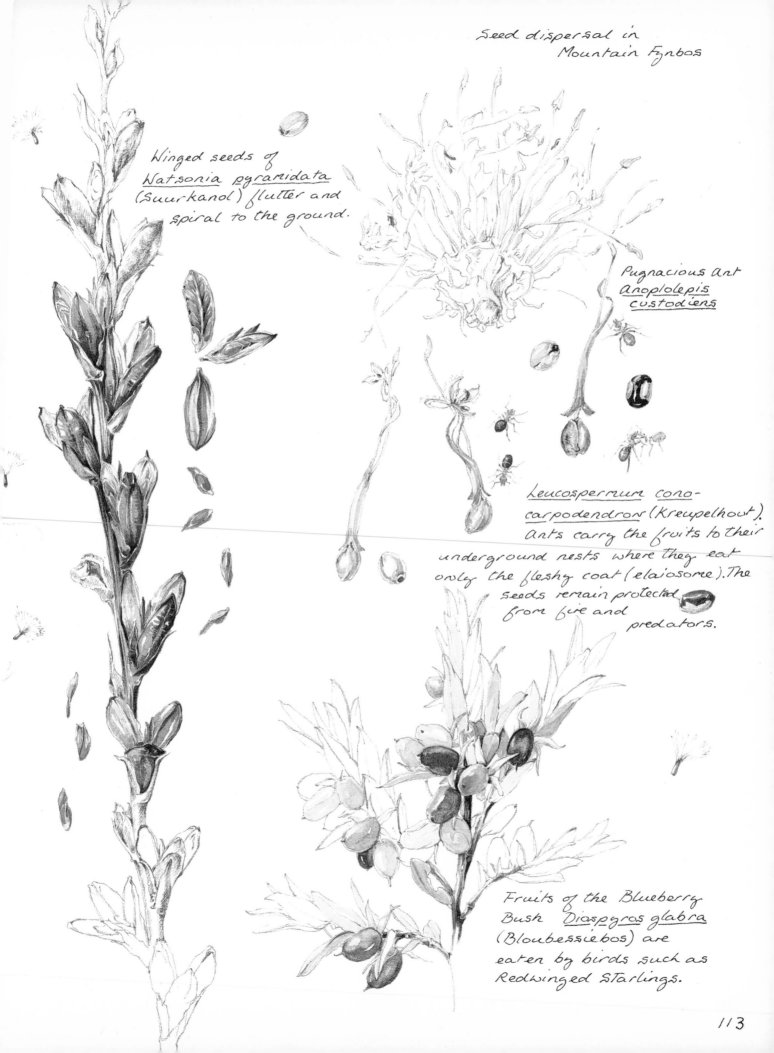

Seed dispersal in
Mountain Fynbos

Winged seeds of
*Watsonia pyramidata*
(Suurkanol) flutter and
spiral to the ground.

Pugnacious Ant
*Anoplolepis*
*custodiens*

*Leucospermum cono-*
*carpodendron* (Kreupelhout).
Ants carry the fruits to their
underground nests where they eat
only the fleshy coat (elaiosome). The
seeds remain protected
from fire and
predators.

Fruits of the Blueberry
Bush *Diospyros glabra*
(Bloubessiebos) are
eaten by birds such as
Redwinged Starlings.

113

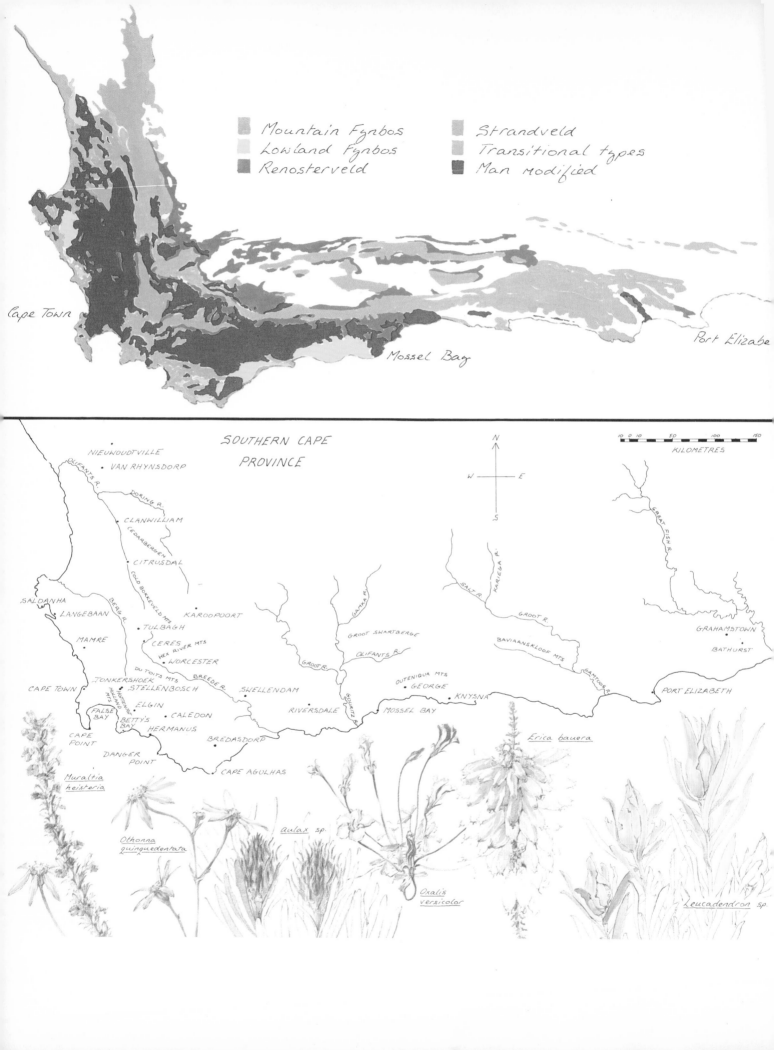

**Mountain Fynbos**
**Lowland Fynbos**
**Renosterveld**
**Strandveld**
**Transitional types**
**Man modified**

Cape Town

Mossel Bay

Port Elizabe

SOUTHERN CAPE
PROVINCE

N
W · E
S

10 0 10    50    100    150
KILOMETRES

NIEUWOUDTVILLE
VAN RHYNSDORP
OLIFANTS R.
DORING R.
CLANWILLIAM
CEDARBERGEN
CITRUSDAL
COLD BOKKEVELD MTS
SALDANHA
LANGEBAAN
BERG R.
KAROOPOORT
TULBAGH
MAMRE
CERES
HEX RIVER MTS
WORCESTER
DU TOITS MTS
BREEDE R.
JONKERSHOEK
STELLENBOSCH
CAPE TOWN
HOTTENTOT HOLLAND MTS
ELGIN
SWELLENDAM
FALSE BAY
BETTY'S BAY
CALEDON
HERMANUS
BREDASDORP
RIVERSDALE
GROOT R.
GOURITZ R.
OLIFANTS R.
GROOT SWARTBERGE
GAMKA R.
OUTENIQUA MTS
GEORGE
KNYSNA
MOSSEL BAY
SALT R.
KARIEGA R.
GROOT R.
BAVIAANSKLOOF MTS
GAMTOOS R.
GREAT FISH R.
GRAHAMSTOWN
BATHURST
PORT ELIZABETH
CAPE POINT
DANGER POINT
CAPE AGULHAS

*Muraltia heisteria*

*Othonna quinquedentata*

*Aulax* sp.

*Oxalis versicolor*

*Erica bauera*

*Leucadendron* sp.

# Fynbos past

## Fynbos beginnings

The present fauna and flora of the Fynbos Biome are the products of evolution and dispersal over millions of years. Long before the appearance of plants and animals, the landscape and soils of the region were forming through movements of the earth's crust, fluctuating sea levels, and the effects of weathering and erosion in a constantly changing climate.

Some 900 million years ago, the continents of the southern hemisphere comprised a single, massive landmass known as Gondwanaland. What is now the southern Cape then lay beneath the sea beyond the edge of this super-continent. Here the accumulation of marine sand and mud formed beds of sedimentary rocks, notably the shales which are now part of the Malmesbury Group.

Between 610 and 500 million years ago, volcanic activity forced an estimated 200 000km$^3$ of magma onto the surface of southern Africa. In the southern Cape these outpourings intruded into and over the Malmesbury sedimentary rocks, where they cooled and solidified to form Cape Granite. These intrusions vary in thickness and texture and, where exposed, form a landscape typified by the rounded granitic domes of Paarl Mountain.

The region then rose slowly above the sea, was transformed into a flat plain by 50 million years of erosion and weathering, and then sank again below a shallow sea. For 200 million years marine sediments settled and accumulated on top of the plain, in places to a depth of 8 000m. About fourteen different geological formations, including shales, mudstones, siltstones and sandstones, were thus laid down at this time. Although collectively termed the Table Mountain Group, these comprise the most extensive sedimentary rock formations in the southern Cape and are found virtually throughout the Fynbos Biome.

Two further sedimentary groups were laid down successively at this time which, together with the Table Mountain Group, comprise the Cape Supergroup. Sediments washed into the sea from the adjacent continent formed the Bokkeveld Group, while the Witteberg Group sandstones and shales were the latest additions to the Cape Supergroup. The Witteberg Group contains rocks displaying evidence of glacial action as Gondwanaland was engulfed by an ice-sheet 1 000m thick about 350 million years ago.

The Cape Supergroup formations were buckled and forced upwards by massive earth movements between 278 million and 215 million years ago, forming the rugged Cape Folded Belt mountains. Although subsequently much modified by faulting and erosion, these form the basis of the landscape that characterises upland areas of the Fynbos Biome today. Table Mountain remained largely unmodified, however, retaining its flat top and stratified layers as evidence of its sedimentary origins. Its lower 70m comprises sandy shales of the Graafwater Formation (a sandstone of the Table Mountain Group), topped by 700m of Peninsula Formation quartzitic sandstones, the whole resting on a foundation of Malmesbury Shales and intrusive Cape Granite.

The Gondwana landmass began to split into Antarctic–Australian and African–South American continental plates some 140 million years ago. The Cape was then positioned about 50–55° South (Cape Town now lies at 34° South) and, with a further 40 million years of continental drift, became bounded by open ocean with the final separation of Africa and South America.

By 65 million years ago, the region looked basically as it does today, but was still 10° south of its present position. The mountains, particularly, with their erosion-resistant rocks, have altered relatively little. The coastal lowlands, by contrast, have been subjected to fluctuations in sea level and the erosion and deposition that accompanied these events.

## The rise and fall of the sea

Over the last 60 million years the sea level has undergone major changes less than ten times, but up to 200m at a time. Thus, the fynbos lowlands have been repeatedly transformed from a virtually non-existent area of dry land to one three times the present size.

Over the last two million years the sea level has undergone more frequent fluctuations because of the freezing and thawing of the polar icecaps. These fluctuations have occurred on a short cycle (relative to those caused by continental tilting and plate tectonics), and have involved changes in sea level of between 6m and 120m. While 6m may not seem very much, it is certainly enough to make you wish your cottage at Betty's Bay was not quite so handy for the beach.

Changes in sea level affect not only the area of exposed land, but the nature of its rocks and soils. The alternating marine and non-marine sediments are mixed with typical coastal deposits eminating from dunes, lagoons, tidal flats and estuaries. The Coastal Foreland of the Fynbos Biome is thus a mosaic of soil types, each with its own peculiar constituents and history, submarine or otherwise. Under these idiosyncratic conditions, fynbos plants and animals have evolved, adapted and alternately colonised and been extirpated. Compared to the relatively stable and ancient environment of the fynbos mountains, the lowlands have provided insecure and changeable circumstances under which successive vegetation types have waxed and waned.

## Past climate

Following a trend over the last 60 million years towards cooler temperatures, the seasonal 'Mediterranean-type' climate which now typifies the Fynbos Biome was established 5 million years ago. A stabilisation of the South Atlantic high-pressure system 2,5 million years ago reinforced the pattern of dry summers in the southern Cape.

The climate of the most recent 2 million years, however, has been dictated by perturbations of the earth's orbit relative to the sun, producing cyclical climatic patterns approximately once every 100 000 years. This periodicity has been characterised by long, cool glaciated spells interrupted by warmer periods of about 15 000 years' duration. There have been between seven and eight glaciations in the last 700 000 years, during which period the Cape coast was cooler by about 5° and wetter than it is now, with rainless spells lasting only three or four days. Snow may have lain permanently at altitudes above 1 000m and occurred as low as 200m above sea level in winter. In fact, the fynbos climate then more resembled that of present-day Marion Island.

We are presently enjoying an inter-glacial warm period which is, nevertheless, approaching its end. But there is no need to start knitting jerseys or laying-in firewood immediately, as the weather will probably not turn colder for a few thousand years yet.

# A history of fynbos flora and fauna

## Early flora

The Fynbos Biome has been described as supporting the most bizarre relict plants in the world, for it contains floral elements which date back millions of years.

Flowering plants first appear in fossil records of the Cape 110 million years ago, when areas with their own distinct microflora were already recognisable in the subtropical climate of southern Gondwanaland. With its separation from South America at this time, Africa existed as an island in glorious isolation until, after 20 million years of drift, it collided with Eurasia. Interchange of plants and animals between Africa and Eurasia, however, seems to have been significant only from about 23 million years ago; moreover, a band of tropical forest across central Africa prevented southward migration. The direct ancestors of southern African flora derive, therefore, from the ancient continent of Gondwanaland. The subsequent development of this flora has taken place largely in isolation.

Some 60 million years ago the fynbos lowlands were clad in a forest of palms, euphorbias and yellowwoods. Fossil pollens indicate an alternation between this tropical vegetation type and a more temperate one, due to climatic fluctuations over the next 10 million years or so.

With decreasing rainfall over the following 40 million years, an expansion of savanna took place at the expense of forest in southern Africa, although evergreen forest probably clothed much of the fynbos region as recently as 5 million years ago, but dying out thereafter. The Knysna forest is a relict of such vegetation, and similar forests still occur at roughly the same latitudes in New Zealand, Tasmania and southern Chile. Seasonally dry Cape sites were subsequently colonised by the drought-resistant and fire-tolerant shrubland vegetation which now characterises fynbos. Reduced leaf size (to minimise water loss) and the development of underground storage organs (to allow resprouting after fire) are common local plant attributes.

The present distributions and abundances of fynbos plants are a response to current dry interglacial conditions. It is appropriate to mention that the vegetation of the northern hemisphere was virtually wiped out in the last Ice Age (which ended 10 000 years ago) when glaciers smothered most of Europe, Russia and North America. The plants found there now are very recent colonists from the south. By contrast, Mountain Fynbos vegetation has not had to contend with such catastrophic extinctions, thereby allowing the uninterrupted development and speciation of its plants. Great antiquity, combined with the need to adapt to micro-environmental conditions (variability in climate and soil nutrients, in particular, over very short distances) together contribute to the exceptional number of plant species within the Fynbos Biome.

## Early fauna

Mammals made their appearance in the fossil record over 200 million years ago. Fossil remains linking them with the reptiles from which they evolved were first found in 1838, near the Karoo town of Fort Beaufort. Such ancient fossils have not been discovered in fynbos, however. Indeed, a picture of the vertebrates that occupied the biome can only be derived from relatively youthful fossils dating back five million years.

Much insight into the early mammalian inhabitants of fynbos has come from the large number of fossils uncovered at a phosphate mine at Langebaanweg. Here, the present-day Berg River once had its estuary, and animal remains found there include those of species that lived on the estuary and others that were washed downstream.

Palaeontological research invariably requires the detective's prowess to piece together clues to the past character of the environment. Seabird, seal and mollusc remains from Langebaanweg point to a sea temperature similar to that of today (a cool 8–14°). Dental abnormalities in fossil teeth suggest that conditions for woodland species were not ideal, and the occurrence of burnt specimens provides evidence of fires. Concentrations of some fossils suggest heavy periodic flooding in which animals drowned and were washed downstream. Taken together, these phenomena indicate a climate that included heavy rain, a dry season and fires caused by lightning.

Many of the fossil remains dating from this period are of animals which have since become extinct, such as a giant three-toed horse, a carnivorous bear, sabre-toothed cats and a four-tusked elephant. Grazers that attest to the grasslands which existed in fynbos at that time are a buffalo, two Bontebok-like antelope, and a gazelle. Large herbivores included an ancestor of today's Squarelipped (White) Rhinoceros. Giraffes are indicative of the presence of trees, but the majority of fossils otherwise point to a predominance of open country in which grasses were an important element.

Two million years ago, open country grazers were even better represented in the fossil fauna of Lange-baanweg as a further decline in forests and woodland had taken place. The Hooklipped (Black) Rhinoceros made its first appearance at this time, as did savanna carnivores such as Lion, Spotted Hyaena and Wild Dog.

More recently, a rich fossil deposit in a Brown Hyaena's lair at Swartklip on the Cape Flats coast has revealed something of life there about 100 000 years ago. False Bay's sea floor was then dry land which, together with the Cape Flats, formed a wide, grassy plain, well watered with streams and vleis. The hyaena lair contained the accumulated bones of herbivorous animals which have since disappeared from the area. Two species, a giant longhorned buffalo and a zebra, have become extinct altogether.

Excavations at Nelson Cave near Plettenberg Bay have revealed the remains of large mammals hunted by Later Stone Age people, when the Agulhas Plain was open grassland and not, as it is now, under the sea. These remains comprise predominantly grazing species including Quagga, Warthog, Black Wildebeest and a form of springbok. This fauna persisted until about 12 000 years ago when it began to be replaced by that encountered by European colonists in the seventeenth century.

## Human history

### Stone Age Man

Homo erectus, the ancestor of modern man, dispersed south from the shrinking forests of tropical Africa about 700 000 years ago and, at some point after this, arrived at and settled in the fynbos region.

By this stage in his evolution, man was already using crude implements in the gathering and processing of his food. The discovery of stone tools, such as hand-axes and cleavers, near Stellenbosch in the last century gave rise to the term 'Stellenbosch Culture' (now known as Acheulian) which labelled these Early Stone Age artefacts. Such tools and other evidence of habitation at this time are widespread in the Fynbos Biome, the majority from valleys and coastal areas and all near sources of fresh water. Excavations in the south-western Cape lowlands have yielded many interesting items from this period and later. Animal remains indicate that the people were hunters or scavengers or both, and there is evidence that they had already

mastered the art of fire-making.

Subsequently, man seemed to become less tied to water in the fynbos region, and his habitations were located higher in the mountains. Artefacts from such Middle Stone Age sites (30 000 to 125 000 years ago) suggest that the occupants had vessels to transport food and water, and made heavy spears as well as cutting, flensing and other working tools. Domestic hearths are evidence of the use of fire 'in the home', and fire was probably also used in the management of game pastures, in game-driving, honey-hunting and in crude agricultural practices such as stimulating the growth of geophytic plants for food. Man's modification of the fire regime and his effects on the vegetation of the Fynbos Biome have certainly been felt for at least 100 000 years.

The continued development and spread of man in fynbos was halted by a cooling and drying of the climate 30 000 years ago. Such conditions led to a fragmentation and isolation of the human population, which are thought to have led to the evolution of the Khoisan from a generalised ancestral negroid stock. The Khoisan were hunter–gatherers, although the relative importance of hunting and gathering as methods of obtaining food is difficult to determine. Hunting of herd animals would have required wide-ranging mobility in the tracking and pursuit of their prey, whereas plant-gathering was a more sedentary pursuit.

The hunting of large herbivores persisted until about 12 000 years ago. By this time, the sea had inundated the coastal plains where most of the animals were to be found. The hunters were forced to adapt to their changing circumstances by broadening their diet. Solitary grazing animals became their prey and greater importance was placed on the seasonal exploitation of underground bulbs and corms. Rock Dassies and tortoises became a major source of protein. At the coast, diet could be supplemented with fish, shellfish and stranded animals. The nutrient-poor soils appear to have thwarted any attempts to domesticate wild plants, however.

### The advent of agriculture

The remains of domestic sheep dating from 1 500 to 2 000 years ago have been discovered in coastal middens east of Cape Town and are evidence of the first farming practices in the fynbos region. Pastoralism in the Cape is thought to have resulted from a westward and southward movement from northern Botswana of Khoikhoi-type peoples, encouraged by the moister climate which prevailed at the time, giving better grazing for their flocks. Competition for resources between the long-established hunter–gatherers and the newly arrived pastoralists was probably fierce. Rock paintings have been found which graphically display skirmishes between the two populations. Certainly the introduction of domestic stock marked the beginning of one of the major impacts on fynbos ecology, that of agriculture.

Herder settlements established over much of the Fynbos Biome were evident until historic times. Such settlements are likely to have had a particularly important effect on the Coastal Foreland where they were concentrated. For example, an estimated 17 000–18 000 Khoi lived between Table Bay and the Berg River in the mid-seventeenth century. A typical temporary encampment at Table Bay, as recorded by the first European colonists, comprised eight or nine huts accommodating 70–80 people who had 700 or 800 cattle and 1 500 sheep. The Khoi were highly mobile, lived in collapsible mat houses and used oxen for transport. The veld was heavily grazed for a few weeks before the animals were moved on. Such a non-selective, heavy-grazing pattern is, in fact, less damaging than constant utilisation, and allows a higher carrying-capacity of stock.

## European colonisation

Jan van Riebeeck arrived at the Cape from Holland in April 1652 to establish a victualling station and fort on behalf of the Dutch East India Company. Cape Town, as this station was to become, formed the nucleus of a rapidly expanding white settlement which went hand in hand with a similarly rapid disappearance of the indigenous people. The population expansion and the consequent exploitation of fynbos resources, have profoundly altered the natural environment. Suffice it to say, at this stage, that in 1687 the resident population of Cape Town was 882. In the 1980s the city and its suburbs support well over 1,5 million people.

# Fynbos discovery

Some appreciation of the complex structure and functioning of the Fynbos Biome could be said to have begun with the activities of the indigenous pastoralists who managed the veld using fire to improve stock grazing. There is also evidence that early hunter–gatherers burnt mountain vegetation to increase yields of *Watsonia* bulbs, which they harvested and ate.

Systematic investigation of fynbos had its beginnings with the succession of enthusiastic naturalists, particularly botanists, who were among the early arrivals from Europe. This period saw the first comprehensive descriptions of fynbos plants, new species of which the disbelieving naturalists seemed to discover at virtually every step. The identities of many of these pioneering naturalists are perpetuated in the multitude of plants and animals named after them.

The first fynbos plant specimen reached the attention of European botanists as early as 1605, when a dried inflorescence of the Blue Sugarbush was illustrated in a book by the Dutch botanist Charles de l'Ecluse (1526–1609). Referred to as 'an elegant thistle' and as having been collected in 1597 in Madagascar, where there are no *Proteas*, this specimen was almost certainly collected near Table Bay where Dutch trading ships replenished on their return voyage from Java. Despite its somewhat unsteady identity and origin, this *Protea* represented the introduction of the Cape flora to the world-at-large.

Encouraged by de l'Ecluse, the Dutch East India Company requested their ships' captains to collect plants on their travels. One such plant, a Chincherinchee bulb which flowered in Amsterdam and was illustrated by de l'Ecluse, had been collected by seamen at 'that extreme and celebrated Promontory of Aethopia commonly called the Cape of Good Hope'.

In subsequent years the number of Cape plants reaching Holland increased, and some of the bulbs commanded extremely high prices from enthusiastic horticulturists. Such was the interest in ornamental plants in the early seventeenth century that the first nursery catalogues were issued at this time. In them, many more Cape species were illustrated, and it was the artists, as much as the collectors, who are remembered for their contributions to establishing the reputation of the Cape's flora.

The first direct recorded observations of Cape plants were made in 1624 by Justus Heurnius (1587–?1653), a missionary returning from Batavia to Holland. His small portfolio included a drawing of a curious and foul-scented carrion flower *Stapelia*. Paul Hermann (1646–95) was the first professional botanist to reach the Cape, en route to Ceylon from Holland in 1672. As a result of his efforts, together with those of a small number of other collectors, almost a thousand species of fynbos plants were described by the end of the century.

Although he never visited the Cape, the great Swedish taxonomist Carl von Linné or Linnaeus (1707–78) played a major role in the advancement of our knowledge of fynbos fauna and flora. Through his contacts in the Swedish East India Company (founded in 1731) he was able to describe and name for the first time many specimens sent from the Cape in the eighteenth century. He was clearly very impressed with what he received. One of his major suppliers was the then Governor of the Cape, Rijk Tulbagh, after whom a genus of lilies, a butterfly *and* a village are named! To him Linnaeus wrote:

May you be fully aware of your fortunate lot in being permitted by the Supreme Disposer of events to inhabit, but also to enjoy the sovereign control of that Paradise on Earth, the Cape of Good Hope. . . . Certainly if I were at liberty to change my fortune for that of Alexander the Great, or of Solomon, Croesus or Tulbagh, I should without hesitation prefer the latter.

No mean praise.

In April 1772, two Swedish naturalists, Anders Sparrman (1748–1820) and Carl Thunberg (1743–1828), arrived in Cape Town. 'At almost every step we made one or two new discoveries,' wrote Sparrman on his arrival. Thunberg, a pupil of Linnaeus, was the first to collect fynbos plants and animals beyond the immediate vicinity of Cape Town.

Plant-collecting was not always the tame pursuit that it may appear. Armed only with a knife and a pair of scissors, Sparrman was confronted by a Hippopotamus (apparently docile, as we have no record of how many snips it took to slay the beast) near what is now Rondebosch; and Thunberg was forced to retreat up a tree in the Knysna forest as a buffalo gored to death three of his horses. After a few months collecting in the southern Cape, Sparrman left to accompany Captain Cook's second expedition in the *Resolution*. He returned to the Cape in March 1775 and continued botanising there until his return in 1776 to Sweden, to which Thunberg, after a trip to Japan, had returned the previous year.

The many scientific publications and accounts of new species that these two Swedes authored are evidence of their achievements. Indeed, Thunberg has been recognised as the 'father of Cape botany' on the basis of his major descriptive works: the two-part *Prodromus Plantarum Capensium* (1794, 1800) and his *Flora Capensis* (1807–20). Their interests were certainly not confined to plants, as Thunberg's collection of over 25 000 insects and numerous birds, mammals and reptiles testifies. Among Sparrman's discoveries was the Cape Siskin, a bird endemic to fynbos, which he collected in 'Hottentot Country' and described in 1768.

A third notable botanist of this period was Francis Masson (1741–1805), a Scot sent to Cape Town in 1772 on behalf of the Royal Botanic Gardens at Kew. His first sojourn lasted 32 months, and he returned to the Cape again in 1785, where he remained for ten years. In addition to collecting specimens, Masson also established a garden in Cape Town. Here, good-quality plants and bulbs could be kept before being freighted back to London. Altogether 400 of the plant species he collected were new to science. These greatly contributed to the pre-eminence which Kew achieved in contemporary botanical circles. Some fifty of Masson's new species were *Pelargoniums*, which, under their popular name 'Geraniums', were soon to become much in demand for horticulture. The genus *Cineraria* was another of Masson's introductions to the world's gardeners. His collection of *Erica* heaths also initiated a passion for these fynbos specialities among horticulturists. One heath, *Erica massonii*, is named after him.

The richness of the Cape flora continued to impress visiting botanists in the nineteenth century. In 1810, the Englishman William Burchell (1781–1863) was astonished to record 105 species in a short walk around Lion's Head, which flanks Table Mountain. This must have inspired him greatly, and over the next four years he travelled 7 000km by ox-wagon, returning from the interior along the southern Cape coast. His specimen collection included over 60 000 plants and 10 000 skins and skeletons.

William Harvey (1811–66), an Irishman appointed to the post of Colonial Treasurer-General in 1835, was responsible for another botanical book of substance to be written on Cape plants. Even before he arrived at Cape Town, Harvey intended to prepare a *Flora Capensis* to 'occupy my leisure', but owing to ill-health his stay was little more than four years. On the basis of his own experiences and the (by then) substantial collections of South African plants in Europe, however, he embarked on the *Flora Capensis: Being a Systematic Description of the Plants of the Cape Colony, Caffraria and Port Natal*. Three volumes were published by 1865, the year before his death. The scale of Harvey's project can be appreciated by noting that the work was completed only seventy years later. Volume IV, edited by Sir William Dyer, a former director of Kew, itself ran to six parts! By bringing together and collating the myriad descriptions of plants scattered through often obscure publications, and by compiling identification keys, Harvey made an invaluable contribution to South African botany.

At the instigation of Sir William Hooker (the Director of Kew), the post of Colonial Botanist was created in 1858 'to perfect our knowledge of the flora of South Africa'. The first incumbent was Dr Ludwig Pappe (1803–62) who was simultaneously appointed to the Chair of Botany at the South African College (now the University of Cape Town). Botanical teaching at the Cape did not have an auspicious start, with most of the students apparently playing truant, and in 1866 the post was abolished. The Chair was reoccupied when Professor Peter MacOwan accepted the position on his appointment as Curator of the Botanic Gardens in Cape Town in 1880. That women were allowed at this date to study at the College for the first time ever is noteworthy! In 1902 the Chair was firmly established, and in 1903 the incumbent was Henry Pearson (1870–1916), who is also remembered as the founder of the National Botanic Gardens at Kirstenbosch in 1913.

Another distinguished botanist of this time was Harry Bolus (1834–1911), a successful financier despite, apparently, spending most of his time in the veld. He collected many thousands of specimens, one half of which were deposited at Kew, and the other half deposited on the sea bottom after a shipwreck off Table Bay in 1879. Bolus published *Orchids of the Cape Peninsula* in 1882 and *Orchids of South Africa* between 1893 and 1911, the year he died. Bolus's contribution is also marked by his founding of a Chair of Botany that now bears his name, at the (then) South African College. He was perhaps the first person to appreciate the unique botanical identity of the southwestern Cape, but it was not until Dr Rudolf Marloth (1855–1931) coined the term *Cape Floral Kingdom* that the region was officially 'christened' botanically.

Marloth's name features prominently in any history of fynbos botany. A German analytical chemist, he arrived in Cape Town in 1883 and was busy collecting plants for his herbarium the day after his arrival. Appointed Professor of Chemistry at Victoria College, Stellenbosch, in 1889, he had as one of his first students Jan Smuts. Marloth combined his interest in botany and mountaineering to great effect. On an ascent of the Matroosberg in the Hex River Range in December 1895 'some 20 species of plant hitherto unknown were found, among them several species of heath, a very showy *Brunia* and some *Lachenalias*'. The author of many scientific papers (including an account of the pollination of the Red Disa by the Pride of Table Mountain Butterfly), his magnum opus was the *Flora of South Africa*, published between 1913 and 1932.

Notwithstanding the long history of fynbos plant-collecting and the lessening chances of finding new species, the enthusiasm of fynbos devotees has not diminished. Notable among recent collectors was T. P. Stokoe (1868–1959), who over 48 years collected some 16 000 plant specimens. Thirty new species were named after him, amongst which was the beautiful *Mimetes stokoei*, now extinct.

That fynbos still holds surprises is illustrated by the discovery of a new, golden *Mimetes* species by a nature conservator, Willie Julies, in the Gamka mountains in June 1988. A new genus of Proteaceae was found as recently as 1984 in the Langeberg near Roberston by Miss Elsie Esterhuysen. Material from the (then) unidentified species was collected as early as 1954, sufficient to arouse suspicion that it was a proteaceous species of unknown generic affinity. It was not until the late 1970s that the site was again visited when, however, it was found that the hillside vegetation had been burnt and only tiny seedlings could be found. Finally, in 1982, the flowers were at last collected and the plant was named by Dr John Rourke of Kirstenbosch as *Vexatorella* (the little troublemaker!) *latebrosa*. For Miss Esterhuysen, formerly of the Bolus Herbarium, this was not her first botanical triumph, as during her long career she has discovered many undescribed plant species from the Cape. Jan Vlok,

working from Saasveld Forestry Research Centre, has a similarly impressive record of finding new fynbos plant species. Perhaps these two botanists, more than anyone else, can appreciate the excitement felt by the naturalists who first encountered the botanical cornucopia of the Cape Floral Kingdom.

The description of new species is only part of the research which can lead to a fuller understanding of fynbos. The current flurry of fynbos research indicates an appreciation of the great scientific and aesthetic attributes of the biome and a realisation that it is under serious threat, particularly from habitat destruction and invasion by alien plants. Jonkershoek and Saasveld Forestry Research Centres, the Universities of Cape Town and Western Cape, and the Plant Protection and Botanical Research Institutes at Stellenbosch, have pioneered and continue much of this research. An additional impetus was provided by the formation, in 1978, of the Fynbos Biome Project, administered by the National Programme for Environmental Sciences of the Council for Scientific and Industrial Research. The results of research are used to assess and elucidate the structure and function of the Fynbos Biome as a whole and the requirements of its constituents. Only on this basis is effective conservation possible.

*Helmeted Guineafowl*

# Fynbos present

## Landscape and soils

Within its narrow band, fynbos extends over a variety of landscapes, ranging from high, subalpine mountain peaks to coastal flats. On the basis of distinctive terrain, geology and soils, the biome can be split into two units – the so-called Cape Folded Belt, which forms spectacular and rugged mountain ranges, and the Coastal Foreland – the gentler plains leading from the foot of the mountains to the sea.

The mountains of the Fynbos Biome are characteristically rocky and steep, the more craggy and precipitous slopes occurring where the rock is more weather-resistant or where there is less rainfall. The Seweweekspoortberg in the Swartberg rises 1 500m in the space of 2,9km, and at 2 325m is the biome's highest mountain. The next tallest peaks are Matroosberg (2 249 m) in the Hex River Mountains and Groot Winterhoek Peak (2 078 m).

Between and on the mountain ranges of the Cape Folded Belt lie valleys and upland plains. Narrow, sediment-filled valleys are typical of the north-west, whereas in the east these tend to be wider. Many of the valleys and plains are fertile agricultural areas.

The wide-ranging environmental conditions which prevail in the Fynbos Biome have resulted in soils which vary greatly in age and nature. On the Cape Folded Ridge soil formation is slow because of the hard, erosion-resistant quartzitic rocks of the Table Mountain and Witteberg Groups which make up its mountains. Weathering takes place continuously, however, and material thus removed from the parent rock generally accumulates in a fan at the mountain bases. Such soils are sandy, very coarse-grained, acidic and very low in nutrients (highly leached). Shallow peaty soils may develop in localised damp, seepy zones.

Soils of the valleys and plains are derived from the ancient shales and mudstones of the Malmesbury and Bokkeveld formations. These soils are heavily textured, clayey (fine-grained) and relatively very fertile compared with those derived from the mountains.

Granite-based soils, although well-leached, are also quite fertile. They are restricted to gently sloping parts of the southwestern corner of the biome and to outcrops on the Cape Peninsula and western forelands.

Foreland soils along the west coast have developed from recent drift sands. Nearest the coast these are highly calcareous and often underlaid by limestone. Inland, there is a reduction in the lime content and a gradual transition to soils similar to those of the Cape Folded Belt quartzitic sandstones. A great variety of soils occurs in the southern foreland, ranging from mountain-like quartzites at Potberg to Bokkeveld-slate derivatives from Bredasdorp to Gansbaai. Limestones dominate between Hermanus and Mossel Bay, with local deposits of acid sands.

## Climate

The Fynbos Biome experiences climatic gradients from north to south, west to east, coast to interior and also with altitude. These gradients, together with the mountainous topography and the influence of two oceans, subject the biome to much local variation in climate. Any generalisation is consequently difficult – what happens at sea level may bear little resemblance to conditions on the mountain tops even a kilometre inland.

Next to fynbos, the greatest attraction of the Cape is sunshine, lots of it. A daily average of about eight hours of bright sunshine may be expected in the western parts around Cape Town, with seven hours on the eastern fringe of the Fynbos Biome at Port Elizabeth. In summer (November to January) bright sunshine in the west generally exceeds a very civilised 10 hours per day, with 7,5 in Port Elizabeth. At this time of year, southwestern Cape weather is regulated by the southern and landward movement of the South Atlantic anticyclone. Long, dry and beautifully warm periods result from this, as does the persistent southeasterly wind, which is such a feature of the Peninsula. This wind, the 'Cape Doctor', is often reinforced by the sea breeze over False Bay, making for windy conditions in the early afternoon.

In fact, the whole coastal belt of the Fynbos Biome is subjected to strong winds. Local peculiarities arise from the juxtaposition of cold oceans and a warm continent. The cooling effect of the sea is especially noticeable in the west where the Benguela current upwells along the Atlantic coast. That this current originates in the cold ocean depths can easily be believed by those hardy enough to take the plunge into its blue waters, whose balmy, tropical prospect belie a distinctly untropical temperature.

Average summer daily temperatures in the Fynbos Biome are about 20° along the south and west coasts but closer to 25° inland in the northwest. In winter it is generally cooler, and the sub-polar westerlies which move north in this season often bring in heavy rain off the Atlantic from that direction. Spectacular squalls can race in from the sea, battering the coast with torrential rain, drops as fat as butterbeans, and driving thousands of albatrosses and other pelagic seabirds close inshore. From time to time, these storms may

*Layers of sedimentary rock forming the cliffs of Table Mountain*

even unceremoniously deposit wayward migrant birds from America on southern Cape shores. But that's another story!

Mountainous areas, where much of the precipitation is orographic, tend to catch the rain much more than the coast or sheltered valleys. Over 3 600mm may fall each year on the Drakenstein Mountains, for example, compared to the 400mm on the Cape Flats or 300mm at Worcester in the Breërivier Valley. The mountains are of great economic importance as their catchments supply fresh water to local urban and industrial consumers.

About 50 per cent of the annual rainfall of the southwest Cape falls in the three winter months (May, June and July). The amount of precipitation is supplemented by moisture from wet stratus cloud which caps the mountain tops in summer. Table Mountain's 'tablecloth' is one example, shrouding the plateau in mist for three months of the year. It has been estimated that such clouds may deposit 500mm of moisture annually. East of Cape Agulhas, the southernmost point of Africa, the Fynbos Biome experiences rainfall at all times of year but only at the eastern extremity is a summer maximum attained. Rainfall along the Coastal Foreland is generally less than 500mm a year, while rainfall of less than 200mm a year is not uncommon in the inland valleys.

Winter snowfalls are fairly regular on the mountain tops, and the sight of the Hottentots Holland dusted with white and brooding beneath leaden skies is certainly impressive. Under such conditions the interior plateau may experience periods of intense cold. Although rare, snow has also been recorded at sea level.

Most disconcerting during the winter months are the berg winds. These hot, desiccating, gusty winds blow from the interior and can raise the temperature by as much as 10° in a few hours.

The southwestern portion of the Fynbos Biome occupies the so-called 'Mediterranean' climate zone of South Africa. Such climate zones, characterised by generally cool, wet winters and warm, dry summers, also occur in California (where the vegetation is known locally as *chaparral*), Chile (*mattoral*), southwestern and southeastern Australia (*kwongan* or *mallee*) and, of course, the Mediterranean Basin (which has many regional names including *batha* in Israel, *garrigue* in France and Italy, *macchia* in Italy and *maquis* in Corsica). Interesting ecological comparisons can be made between the plants of these areas, which show striking similarities in their ability to cope with, for example, prolonged summer drought. Such adaptations-in-common, termed 'convergent evolution', have taken place as a response to similar environmental pressures and regardless of the fact that the regions are very distant from one another. A high degree of plant-species richness and endemism is also found in these other 'Mediterranean' areas, but not one approaches the magnitude of that occurring in fynbos. An added similarity which well merits investigation is the high quality of the wines produced in these regions.

## Fynbos flora

Since the first collections of southern African plants reached Europe in the late sixteenth and early seventeenth centuries, the flora of the Cape has been famous for its richness, beauty and scientific interest. By the eighteenth century, it was appreciated that here was a botanical region strikingly different to any other found

*Additional summer moisture is supplied to the mountain tops by orographic clouds, here spilling over the Hottentots Holland on the east coast of False Bay*

in Africa or, indeed, anywhere else in the world.

With its complement of at least 8 578 species of flowering plant, the Fynbos Biome is now recognised as supporting one of the most diverse and distinctive floras in the world. In one fynbos patch, measuring only 10x10 metres, 121 different plant species have been recorded. This small-scale richness is apparently exceeded in other temperate regions only by Israeli shrublands.

Indeed, on a grander scale, the Fynbos Biome supports the world's richest flora, with 1 300 species per 10 000km², compared to about 400 for its nearest rivals in South American rain forest. The 60km² of Table Mountain alone support about 1 470 species, which is more than the British Isles (308 000km² and 1 443 species). Altogether 2 256 species occur on the Cape Peninsula (470km²).

All in all, 5 832 (68 per cent) of the Fynbos Biome's plant species and 193 (20 per cent) of its 955 plant genera are endemic. Not only are many of the plants restricted to the biome as a whole, but many also have extremely small ranges within the biome. The *world* distribution of *Erica fairii*, for example, is a single hectare of the Cape Peninsula. A large and curious orchid, *Satyrium foliosum*, apparently grows only on two cliff ledges, and another heath, *Erica sociorum*, is confined to a scattering of rocky crevices on a single mountain near Fish Hoek. *Leucadendron macowanii* occurs only in one small gulley south of Simon's Town. The tiny ranges of many fynbos plants have rendered them dangerously susceptible to extinction.

The extraordinarily high plant-species richness of the Fynbos Biome and the unusually small ranges of many of its plants are attributable to features of the landscape and environment. The dissected topography of the region isolates plant populations, favouring local differentiation. Climate, soil nutrients and fire regime vary significantly between sites, enhancing speciation as the plants adapt to their peculiar circumstances. Many fynbos plants have their seeds dispersed by ants. The short distances (relative to wind-dispersal or bird-dispersal) over which the seeds are thus carried also promote the evolution of species through the absence of genetic mixing with other populations.

# Fynbos vegetation types

The plants of the Fynbos Biome together comprise three major vegetation types: Renosterveld, Strandveld and Fynbos. In addition, small patches of Kloof Woodland occur on scree slopes and streambanks. Afromontane Forest, Karroid Shrublands and Kaffrarian Thicket or Valley Bushveld have patchy or peripheral distributions in the biome, but are found widely elsewhere in southern Africa and are not considered further here. The extent and distribution of the Fynbos Biome's major vegetation types are shown on the map.

## Renosterveld

Four types of Renosterveld vegetation have been identified by botanists, but, generally, it is found on fertile, shale soils of the Malmesbury and Bokkeveld Groups in low-lying, gently undulating areas. Before its recent transformation to agricultural land, Renosterveld is thought to have comprised largely *Themeda triandra* grassland. Today, asteraceous species (daisies) are usually abundant, and geophytes (plants with underground storage organs such as bulbs and corms) of the iris, lily and sorrel families are also important components. Small, isolated thickets comprising mainly the Wild Olive and Common Guarri, also typify this vegetation type. Such thickets are often confined to curious circular, gravelly earth mounds up to 20m in diameter and less than a metre high, known as *heuweltjies*. These are thought to be produced by the subterranean activities of molerats. More typically, Renosterveld vegetation is short-leaved and bushy and dominated by *Renosterbos*, the Rhinoceros Bush. It is difficult, these days, to imagine rhinoceroses patrolling the Renosterveld-clad slopes of Cape Town's Signal Hill, but they did so until European settlement of the Cape in the mid-seventeenth century.

## Strandveld

Strandveld – coastal vegetation of deep, well-drained shell sands – is restricted to the southwestern end of the region from the Gouritz River to the Cape Flats and north up the Atlantic coast to the limits of the biome in Namaqualand. It occurs from sea level to about 150m.

Strandveld vegetation is highly variable, its species and structure dependent on local environmental conditions. Hard-leaved evergreen shrubs, such as White Milkwood, Sea Guarri and other berry-bearing shrubs, which attract flocks of frugivorous birds, tend to dominate in old Strandveld. Tall, rush-like restioids occur in scattered patches, with shorter restioids being more generally distributed. Strandveld's spectacular spring-flower display, when the veld is carpeted with daisies and fleshy-leaved *vygies* (mesembryanthemums), is a sight not to be missed.

## Fynbos

*A Fynbos Year* concentrates primarily on the most extensive vegetation type in the biome – Fynbos itself.

Fynbos vegetation is generally characterised by a scarcity of trees, few grasses or evergreen succulents, and a prevalence of shrubs adapted (through small, incurled, or hard, evergreen leaves) to withstand water- and nutrient-shortage. It is notable for its high species richness and concentration of endemic plants (Renosterveld and Strandveld, by contrast, support few endemics).

Comprising Mountain and Lowland Fynbos, it occurs on relatively nutrient-poor sandstone and quartzitic soils of the Cape Supergroup (Mountain Fynbos) and sand plains and limestone (Lowland). Mountain Fynbos vegetation is typical of many of the region's most beautiful areas, the Cedarberg, Hottentots Holland and Outeniqua mountains, for example. Paradoxically, perhaps, Mountain Fynbos also occurs at sea level at the Cape of Good Hope Nature Reserve and other sites where environmental conditions resemble those of the mountains.

Lowland Fynbos occurs on the dunes and rolling hills along the western and southern Coastal Forelands from sea level to 150m above sea level. It may be categorised according to substrate (whether deep acid sands, calcareous shallow sands or limestone). The limestone flats of the Agulhas Plain support a particularly species-rich Lowland Fynbos plant community.

## Fynbos trees and Kloof Woodland

There are few trees in Fynbos, a phenomenon that has been attributed by some botanists to too-frequent burning and over-exploitation in modern times. Alternatively, such trees as do remain may be relicts of larger forests whose remnants are now restricted by climatic change.

The Clanwilliam Cedar (endemic to the Cedarberg) and the smaller Mountain Cypress (p. 37) are two species which persist on open hillsides. Where the vegetation is protected from fire, along riverbanks or on rocky cliffs and scree slopes, woodland patches may occur. Here the vegetation is strikingly different to that of the adjacent Fynbos, and includes tall trees such as yellowwood, Butterspoon, Wild Peach (p. 95), and the pink-flowered *Virgilia oroboides* (p. 20). The Wild Almond is also largely restricted to stream banks, and the Fountain Bush and showy Waterblossom Pea (p. 62) are also typical of this habitat.

## Characteristic Fynbos plants

It is generally accepted that Fynbos is characterised by proteoid (leathery-leaved, woody shrubs), ericoid (heath-like, low, evergreen shrubs with narrow rolled leaves), and restioid (rush-like plants with near-leafless, wiry or thin, hollow stems) components. These plants resemble typical members of the Proteaceae, Ericaceae and Restionaceae, respectively, but are not exclusively members of them. They do, however, typify much of Fynbos vegetation and are probably its most familiar botanical occupants.

## Proteaceae

Preponderant amongst the proteoid component is the family Proteaceae. Its members are probably the best known and certainly amongst the most enigmatic, evocative and showy plants to be found in Fynbos.

The Proteaceae is apparently one of the oldest families of flowering plants. Before the break-up of Gondwanaland it had divided into two subfamilies, the Proteoideae and Grevilleoidae. Southern Africa now supports most members of the former, while the latter occurs in Australia and South America. Many proteaceous genera are shared between Australia and South America but none is found in both Australia and South Africa. Only one South African species, the Wild Almond (famous as Van Riebeeck's Kirstenbosch hedge and, as such, the first indigenous plant to be cultivated here) of the southwestern Cape, is closely related to the Australian Proteaceae.

Proteus, after whom this family and the genus *Protea* were titled, was a Greek sea-god, fabled to take on a multitude of shapes and forms. So it is with the plants named after him. The Proteaceae comprises 14 genera including, in South Africa, *Mimetes*, *Leucadendron* (conebushes), *Leucospermum* (pincushions) and *Protea* (sugarbushes). These and their close relatives range in size from prostrate, dwarf creepers to erect, tall bushes or trees (*Faurea macnaughtonii* may be 20m high). The individual flowers are small and inconspicuous, but are grouped in a conspicuous inflorescence often surrounded by colourful bracts. The leaves are tough and leathery, broad or needle-shaped, and in some cases (*Leucadendron* species, for example; p. 41) are brilliantly coloured.

Although members of the Proteaceae occur north to the Limpopo River, over 90 per cent of the family's more than 330 species are found only in the Fynbos Biome. Here, many also have tiny natural ranges. The exquisite Blushing Bride, for example, is confined to Assegaaiboskloof in the Franschhoek Mountains. Once thought to be extinct, it reappeared when dormant seeds germinated after a wildfire. It is now a justifiably popular garden shrub and cut-flower species (p. 35).

There is not a month when some member of this family is not in flower. Winter in the southwestern Cape, with its copious rains, sees the blooming of many species whose bright displays provide welcome respite from grey days and gumboots. A mountain slope clad in flowering sugarbushes, replete with prattling Cape Sugarbirds and darting sunbirds, is certainly one of the most splendid sights which Fynbos has to offer. Apart from their aesthetic qualities, the birds may fulfil an important role by pollinating the flowers. Insects (such as Cetoniid beetles, p. 16), also visit the inflorescences, especially on warm days.

The relative importance of avian and insect pollination is a subject of some scientific contention, but it may be that the transference of the pollen is, in fact, effected by tiny flower mites. These are present in huge numbers in *Protea* blooms, in particular, and hitch lifts between plants on the facial feathering of the bird visitors or on the insects. Pollen grains adhering to these mites may become dislodged as the latter disembark into the inflorescences. As the birds and insects are necessary for transporting the mites from plant to plant, their role is indispensable, regardless of whether they directly pollinate or not. Beyond such controversy are ten *Leucadendron* species and all three members of the genus *Aulax* which are wind-pollinated.

About a dozen species of Proteaceae are illustrated here, but do remember that, conspicuous as many of them may be, members of this family comprise a mere 5 per cent of the plant species that can be found in Fynbos.

## Ericaceae

Perhaps the next best-known family of Fynbos plants is the Ericaceae or heaths. The name derives from the Greek *ereiko*, to break, as a herbal drink made from their leaves was reputed to break or dissolve gallstones.

Fynbos certainly maintains a monopoly of this attractive group, with over 670 species, 650 of which are endemic. In the Cape Peninsula 103 species occur, with an impressive 220 in the mountains around Worcester and Caledon. Elsewhere in the world, there are 14 *Erica* species in Europe, 9 in Malawi and northeast Africa, and another 2 in North Africa. The Fire Heath (p. 53) has the widest distribution of South Africa's *Ericas*, but the majority have much smaller natural ranges.

*Ericas* may grow as high as 6m but all have the same short, narrow leaves. The flowers are more variable, ranging from pinhead-size to 6cm long. The basic shape of the flower is an open or closed bell and they may be smooth and soft, hard and shiny or even sticky or hairy. Most are pink, but various shades of green, yellow, orange or red and combinations of these also occur.

Such variation in flower form reflects the pollination mechanism of the particular *Erica*. Wind-pollinated species have small, relatively inconspicuous flowers which release a cloud of pollen when shaken. Insect-pollinated *Ericas* (which account for 80 per cent of the species in the southwest Cape) have larger and brighter flowers than wind-pollinated ones, and often display adaptations which facilitate visitation by certain types of insect. The length of the flower tube of some species, for example, is perfectly matched by that of the proboscis of the flies that visit them. Although few *Ericas* are scented, there are a small number of moth-pollinated species which smell strongly at night.

Over 60 *Erica* species in the southwest Cape appear to be bird-pollinated. This is perhaps surprising, as only one bird species, the Orangebreasted Sunbird (p. 95) has the beak morphology and geographical distribution necessary to make it a dependable *Erica* visitor. Nevertheless, such *Ericas* have strongly curved, tubular flowers which conform to the length and shape of the sunbird's beak, and produce copious energy-rich nectar which the birds drink. Others have straighter flowers with a barrage of bracts and sepals or stickiness and hairs on the flowers to deter nectar-stealing insects. These unwelcome guests 'short-circuit' the pollinating mechanism by piercing the flower tube at its base to gain access to the nectar. Bird-pollinated *Ericas* also have significantly thicker stems than wind-pollinated or insect-pollinated species, apparently to support the extra weight of their visitors.

Some *Erica* species are illustrated on pages 8, 12 and 40.

## Restionaceae

The third major group of plants that distinguishes Fynbos is the Restionaceae. This family includes twelve genera, all reed- or rush-like, comprising 310 species, 290 of which are endemic to the biome. The male and female flowers are carried on separate plants, which may give the illusion that two different species are involved. Their identification to species level, if not completely impossible, is at least fiendishly difficult. That, at any rate, is our excuse for the page of almost anonymous spikes (p. 2). Although lacking colourful flowers or vivid foliage, the restios are very attractive in their way, adding a bronzy sheen to the landscape, or a patchwork of subtle greens and browns, restful to the eye.

There is some evidence that restios are adept at exploiting nutrient-poor habitats through their efficient resprouting rooting-system. As burning frequencies have lately increased over much of the biome, the nutrient loss that accompanies fires has depleted soil-nutrient levels and rendered many areas suitable for colonisation by restios.

A typical Fynbos scene will, therefore, include proteas, ericas and restios as its basic components, leaving only about 6 000 other plant species to fill the gaps. . . .

## Some other fynbos flowers

Although the family Ericaceae contains the largest number of endemic species in fynbos, a number of other families match or exceed it in richness. For example, 986 species of the daisy family occur in the Fynbos Biome, 608 of them being endemic. The Iridaceae (irises) with 660 species, 485 of them endemic, ranges from the elegant *Gladiolus brevifolius* (p. 30) and *Gladiolus maculatus* (p. 42) to the bizarre *Ferraria crispa* (p. 81) and Rat's Tail (p. 56).

Over 124 Geraniaceae species (including 67 endemics) occur in fynbos. Some are small and quite delicate (such as *Pelargonium myrrhifolium*, p. 57), others are robust, woody shrubs up to 2m in height. Their variety of form and colour and ease of cultivation have made them popular ornamentals, and this family is the source of the familiar 'Geraniums' which now adorn the window-boxes of the world.

A family that merits special mention is the Orchidaceae. Fynbos supports a particularly impressive selection of these enigmatic flowers. Over 200 species occur, ranging from the glamorous Red Disa through the pink-and-white *Satyrium* species (p. 82), to the more demure *Pterygodiums* (p. 69) with their curious soapy scent. The latter are every bit as attractive, in their modest way, as their larger and more demonstrative relations. We must admit to finding the smaller, unpretentious and hard-to-find species much more exciting than the ones that, in a manner of speaking, leap at one out of the bush. Fynbos is full of surprises, so when you are next admiring the regal splendour of a King Protea (illustrated on the title-page) at eye-level, have a good rummage in the tangle at your feet for hidden gems.

It is possible to find plants in flower at any time of year in fynbos, but spring does herald the finest displays, particularly where the vegetation has been burnt the preceding autumn and where the winter rains have encouraged regeneration. The reaction of some plants to fire can be very rapid, however. Within eight days of a burn at Swartboskloof, the first Fire Lily appeared. Tall, pink *Watsonia* flowered in the following months, with bright blue *Aristea*, deep yellow *Wachendorfia* and brilliant yellow, nodding *Bobartia* towering over the many *Gladiolus* species. These, in turn, overcanopied a bewildering collection of more diminutive 'Phoenixiphytes' rising from the ashes – the strange *Wurmbea* (p. 63), star-like *Spiloxene* (p. 62), and exquisite *Roella*.

The last-named were also flowering profusely in an area of the Cape of Good Hope Nature Reserve, two years after a wildfire (p. 29). Here, the combined effects of salt-laden winds and grazing by buck may retard the recovery of the vegetation. Nevertheless, there were some magnificent patches of flowers to be found in the young veld, including bright blue *Lobelia* and Shepherd's Delight with its citrus-scented leaves (p. 56).

To attempt to describe any more than this handful of fynbos flowers would require a good many more volumes. When you leaf through the illustrations, do remember that they portray only the tiniest fraction of the plants that the Cape Floral Kingdom has to offer.

## Fynbos mammals

In contrast to the plants, the mammals of the Fynbos Biome are generally not distinct from those found elsewhere in the subcontinent. About 90 of southern Africa's 280 mammal species have been recorded in fynbos. The exact number of endemic species is uncertain, largely because the distributions and taxonomic affinities of many of them, particularly the smaller ones, are poorly known. Of the small mammals, Van Zyl's Golden Mole (known only from a single specimen collected on the west coast at Lambert's Bay), Duthie's Golden Mole (found only on the coast between Knysna and Port Elizabeth), Cape Dune Molerat, Cape Spiny Mouse, Cape Gerbil and Verreaux's Mouse may be considered to be largely, if not wholly, restricted to fynbos. Of the larger mammals, only the Grysbok and Bontebok (a variety of Blesbok) are endemic.

Fynbos is not a land of large, spectacular mammals, contrary to the expectations of many visitors. This has not necessarily been the case always, however, as the arrival of European colonists in 1652 sealed the fate of those larger species which had survived climatic change and hunting by early man. The patient observer may still find species of interest but, for the most part, must be prepared to be content with mammals no bigger than a small buck or even a mouse!

With the extirpation of the Lion, the mammal predator's crown in fynbos is worn by the Leopard, which is probably more numerous in the mountains of the southern Cape than is generally appreciated. They are seldom seen, however. Our closest encounter was the tantalising discovery of their spoor at Swartboskloof. We wondered just how often one had watched us pass by on the mountain paths.

The handsome Caracal or Lynx, with its rich brown pelage and black-tufted ears, is another important fynbos predator. The predominantly nocturnal Cape Fox has only appeared once for us (at Jonkershoek). The curious-looking Bateared Fox is, sad to say, most commonly encountered as a road casualty. Nevertheless, it appears to have recently expanded its range in the Cape Province, including the Fynbos Biome where it is most common in Strandveld and Renosterveld. Its huge ears (p. 98) are used to detect underground insects, particularly termites and beetle larvae. Lizards, small rodents and even fruit comprise the remainder of its diet.

The Aardwolf and Antbear (Aardvark) are two other large mammals whose modest diets are composed largely of ants and termites. Although their distributions do extend into the Fynbos Biome they are probably not as common here as in other parts of the country. The Antbear is found especially where heavily utilised pastures support large termite populations. Neither is often seen, and our experience of both animals is limited, in fact, to my falling into an Aardvark burrow near Saldanha.

Two genet species, the Small-spotted and the Large-spotted (p. 110), are found in fynbos. These are members of the family Viverridae which also includes the mongooses, civets and Suricate. The genets look, perhaps, like a cross between a dog and a cat, with beautifully marked coats, ringed tails (white-tipped in the Small-spotted, black in the Large-spotted) and long, sleek lines. Insects and mice comprise the bulk of their diet and both species are strictly nocturnal.

Evidence for the presence of Porcupines is often limited to a few shed quills, and signs of digging and rummaging for geophytic plants. The Porcupine will not distinguish between wild and cultivated species and will happily uproot any vegetables or ornamental bulbs you may care to plant for it in your garden.

Other fynbos diggers are the golden moles and molerats. Six species of the former occur in fynbos. All are blind, have powerful forelimbs for tunnelling, sheens of various colours on their coats, and dense underfur. Golden moles feed on insects or earthworms, and they prefer loose soils or light sandy loams in which to burrow. Very little is known about their life-histories or ecology apart from the fact that the Cape Golden Mole can be a nuisance when its activities uproot garden plants. Incidentally, this species was the first of the golden moles to be described scientifically, although in his original account of 1758, Linnaeus was a little off target when he plotted the collecting locality as 'Siberia'!

The molerats (whose family name, Bathyergidae, comes from the Greek *bathys* meaning deep, and *ergo*, to work) are apparently more closely related to the Porcupines than any other mammal. Despite their name, they are neither moles nor rats. A striking feature is the enormous incisor teeth which grow outside the lips so the animal can tunnel and not get a mouthful of soil at every bite. Three of the five South African species are found in the Fynbos Biome, mainly in well-drained alluvial lowland soils in Strandveld and Renosterveld. The largest is the Cape Dune Molerat which can measure up to 33cm long and weigh up to 750g. It is, in fact, the largest obligate burrowing animal in the world.

Molerats feed exclusively on the underground storage organs of geophytic plants, notably those of the families Iridaceae, Liliaceae and Orchidaceae. Recent studies have revealed an interesting relationship between one fynbos geophyte, namely *Micranthus*, and the Cape Molerat. To deter would-be predators many plants accumulate toxic or distasteful compounds in their underground storage organs. The molerat relishes *Micranthus* corms, however, and also transports them along its burrows to store in subterranean larders. This apparent paradox (surely being palatable will not increase the plant's survival chances?) actually enhances the effective distribution of the plant. The *Micranthus* corm is made up of a number of segments, like those of a garlic head, enclosed within a protective tunic. Clusters of small cormlets are loosely attached to the stem above the corm. As it feeds, the molerat peels off the fibres enclosing the whole corm, dropping some of the component segments and dislodging one or two cormlets from the stem in the process. Having removed one corm segment to eat, the molerat drops the parent corm again, dislodging more cormlets and segments, some of which it fails to relocate. In this way individual segments and cormlets are safely dispersed and left to germinate.

This adaptation not only conserves precious nutrients which would otherwise have to be diverted into manufacturing anti-molerat chemicals, but increases the plant's chances of effective dispersal. It was no coincidence that *Micranthus junceus* was the most abundant geophyte at the Lowland Fynbos study site near Darling. If you are going to be eaten, you might as well turn it to your advantage! It is likely that many such intriguing relationships between fynbos plants and animals await discovery.

So far, most of the fynbos mammals we have dealt with are rarely seen because of their nocturnal, subterranean or retiring natures. There are some species that can be seen in daylight, however. One that may require an early rise is the Cape Clawless Otter. This large otter occurs in rivers, dams and coastal waters throughout the biome. Although they may be seen at any time of day, they are most active for the few hours after sunrise and again just before sunset. In our experience, however, distinctive clawless prints in the sand are more often seen than the animals themselves. Crabs and frogs are favoured prey, and latrines containing remains of these are a clear indication of the otter's presence.

Four species of mongoose are found in fynbos, the most commonly observed being the Small (Cape) Grey Mongoose. This diurnal and predominantly insectivorous species will also scavenge in roadside rubbish bins. The Yellow Mongoose is also diurnal, living in underground colonies of 20 or more individuals. If the yellow coat is not always distinctive, the white tip to the tail generally is.

Few buck or antelope occur in fynbos, and very few are herd species such as are found in savanna. The sure-footed Klipspringer is well distributed in Mountain Fynbos and other rocky, upland areas elsewhere in southern Africa. It is a small antelope standing about 60cm high at the shoulder, and occurs in pairs or family groups, feeding mainly by browsing on bushes and shrubs. Two adaptations which equip the Klipspringer for living in a rocky habitat set it apart from other African antelopes. Firstly, by walking on the very tips of its hooves the Klipspringer wears these down to form an oval pad which is more efficient at gripping smooth rock surfaces. Secondly, the hairs are hollow and very thick (like those of the Polar Bear), giving efficient insulation and also, perhaps, cushioning the animal in a fall. The latter attribute was appreciated by the European colonists, who used the hair to stuff their saddles.

Two smaller antelopes found in the biome are the Steenbok and Grysbok. The former is widespread in the subcontinent, but the Grysbok (p. 10) is confined to fynbos. The grey grizzling of its brown coat and its habit of crashing unceremoniously through the undergrowth when disturbed distinguish it from the Steenbok. The latter is a rich rufous brown and tends to make its escape more gracefully by springing and twisting and turning at speed.

With its bright white blaze and purple-glossed brown coat, the Bontebok is a handsome antelope. It is now considered to be a subspecies of the Blesbok. The Bontebok is today found only in a scattering of game farms and at the Bontebok National Park at Swellendam, De Hoop Nature Reserve and the Cape of Good Hope Nature Reserve. Bontebok characteristically pose on warm days, facing the sun with head slightly lowered and occasionally snorting, foot-stamping or head-shaking.

Another relatively conspicuous, but much more widely distributed mammal is the Grey Rhebok, an elegant antelope of rocky mountain slopes and mountain plateaux. The name seems to be a corruption of the English 'Roebuck', which the Rhebok, with a little imagination, superficially resembles. The scientific name of the Roe Deer, *Capreolus capreolus*, is the origin of the Grey Rhebok's scientific epithet (*Pelea*) *capreolus*.

The predominance of browsing, rather than grazing, mammals in the Fynbos Biome is indicative of the scrubby and bushy nature of the vegetation. Natural grazing in the western and southwestern Cape has been found to be of very limited food value, with micronutrients such as phosphorus, copper, cobalt and manganese being in very short supply. As a result, the number of large grazing species which fynbos can support tends to be naturally restricted. Small browsing antelopes fare on the whole better, as their stature and relatively small muzzles enable them to select plant parts with the highest food value. Because they eat less than larger species they can spend more time actively seeking the richest and most nutritious food.

For the casual fynbos visitor, the most frequently encountered mammals are probably the Rock Dassie and Chacma Baboon. On a warm day the visitor to the Table Mountain plateau has to vie with dassies for bench-space, possibly unaware that he or she is in the company of a creature whose nearest relatives are the Dugong (manatee) and Elephant. Dassies are small, tailless animals, weighing up to 4,5kg, with yellowish-buff, reddish or greyish-brown fur and an enigmatic smile (p. 83). They are sun-loving, social animals and exclusively vegetarian. At 230 days (almost eight months) the gestation period is long for such a small animal. Within a day of being born the young eat solid food and move confidently over rocks and boulders.

The Chacma Baboon's catholic diet and frequent indifference to man and traffic make it a familiar sight at roadsides and carparks on the Cape Peninsula and elsewhere. Baboons are highly gregarious, living in troops of up to a hundred individuals, with one or more dominant males. Aggressive encounters are frequent, particularly between young males, and the quiet foraging of a troop is often disrupted by unruly and noisy outbursts. More endearing behaviour is displayed by the females when looking after their young and by the obviously pleasurable bouts of mutual grooming indulged in when the animals remove ticks, fleas, salt-flakes and other choice items from each other.

The Fynbos Biome has far more small mammal species than large ones, but, though rather difficult to see, they are every bit as interesting. The most commonly encountered small mammal in our fynbos year was the Striped Mouse (p. 6). The four black stripes on its back make identification of this diurnal and engaging little rodent quite easy. The larger Vlei Rat, also quite conspicuous by day, is found in marshy areas, where small elevated platforms neatly littered with droppings and gnawed restio stems confirm the animal's presence.

The Namaqua Rockmouse is known to play a role in the pollination of ground-flowering *Protea* species. In its efforts to reach the nectar, the mouse nuzzles deep into the inflorescence, picking up pollen on its fur in the process. This pollen may be deposited on the stigmata of the next *Protea* which the mouse visits.

In spite of their modest size, small rodents probably play a disproportionately large role in the fynbos ecosystem. They are important predators of seeds, especially in post-fire vegetation, and in turn are themselves a major source of food for a variety of predators, ranging from snakes to birds of prey, and carnivorous mammals from mongooses to Leopards.

Probably the least known of fynbos mammal groups, but potentially one of its most fascinating, are the bats. To most people, bats are probably just a flutter of wings at dusk, but they are certainly deserving of greater interest and sympathy. Being active only by night and roosting by day in often inaccessible and inhospitable caves do not make the bat amenable to casual study. Some 18 or so species have been recorded in the Fynbos Biome, rejoicing in such evocative names as the Strawcoloured Fruit Bat, the Tomb Bat, the Flatheaded Freetailed Bat, Schreibers' Longfingered Bat, Temminck's Hairy Bat and Geoffroy's Horseshoe Bat.

Most fynbos bats are insectivorous, a notable excep-

tion being the Egyptian Fruit Bat. This wide-eyed, long-winged species, illustrated on page 11, appears to be becoming increasingly rare on account of persecution (it may damage soft-fruit crops) and destruction and disturbance of its roost sites.

The most abundant bat in the region is probably Schreibers' Longfingered Bat, for which one cave at De Hoop Nature Reserve may provide home for over 100 000 individuals. This important roost and maternity site probably contains over 150 000 bats of five species at one time. These consume an estimated 410kg of insects each night! The 30° temperature and high humidity of the cave are ideal for bats, which appear quite selective in their choice of breeding and roosting sites. Research has shown that the females of some insectivorous species at De Hoop migrate to colder regions to hibernate. This allows them to gestate over the winter months, when they would otherwise have to feed themselves and the developing young in a period of insect shortage.

# Fynbos birds

Unlike many other fynbos inhabitants, birds are generally (but not always) conspicuous, relatively easy to identify and, unlike flowers, they come to you (sometimes). Birds are also our favourite fynbos occupants and are responsible for our association with it, so we feel inclined to give them a favourable review.

From the scientist's point of view, birds are the best-known group of animals in fynbos. It has to be said that this is more a reflection of the lack of knowledge of other wildlife, rather than of a particular wealth of information on birds, although the last decade has certainly seen an upswing in fynbos-bird research. Descriptions of the bird communities of various fynbos vegetation types are now available, and aspects of the biology of certain species and groups of species, notably the nectar feeders, are now reasonably well known.

### Endemism and species richness
About 250 species of land and freshwater birds are found in the Fynbos Biome, but in contrast to the plants, the number of endemic species is curiously small. Cape Rockjumper, Victorin's Warbler, Cape Sugarbird, Orangebreasted Sunbird, Cape Siskin and Protea Canary are the only species restricted to the biome.

Among the various theories propounded to account for the relative paucity of endemic fynbos birds are the 'low structural diversity' of the vegetation (particularly the absence of trees), an apparent shortage of seeds suitable for granivorous birds, and a general progressive reduction in the number of species of birds in Africa south from the equator. There is also an idea that, instead of many bird species evolving in fynbos, each occupying one available habitat or niche (as appears to be the case with fynbos plants), a few adaptable birds are able to exploit the majority of them. A simpler and more plausible explanation is that, being very mobile, birds are not subject to the geographical isolation and micro-environmental pressures that have induced high speciation among fynbos plants.

## How many birds in fynbos?

Observations at some of our Mountain Fynbos study sites have almost forced us to conclude that 'fynbos bird' is a contradiction in terms. With less than a single bird per hectare, and a mere nine species recorded in that one hectare over a period of two years, it is tempting to label Mountain Fynbos, at least, as an ornithological desert. We need to give qualification to this ruthless generalisation, however.

Fynbos bird communities vary according to the nature, age and location of the vegetation they occupy. About seven birds per hectare can be found in Mountain Fynbos generally, but in old vegetation where *Proteas* and *Ericas* are flowering, this figure can be considerably higher. Indeed, we have seen an estimated 70 Orangebreasted, 30 Lesser Doublecollared and 10 Malachite Sunbirds, and 5 Cape Sugarbirds in a patch of flowering *Erica gilva* approximately 100m$^2$ near Cape Point.

Studies of Coastal Renosterveld and Strandveld sites have recorded bird densities of some 10 and 12 birds per hectare respectively, with 12 and 30 species recorded overall. In a subcontinental context the bird densities of these vegetation types are on a par with eastern Cape bushveld or Mopane Woodland in the Zambezi Valley, although the bird species involved are, of course, different. Compared to tropical rain-forest, on the other hand, fynbos is, in an ornithological sense, decidedly dull. In Papua New Guinea, 165 species of birds have been recorded in a 2,5ha rain-forest plot, with an average of 169 individual birds per hectare counted on each visit. The mind boggles. If this seems an unfair comparison, it should be remembered that there are often as many plant species per unit area in fynbos as there are in tropical rain-forests, which are popularly held to be richest in just about everything.

## Fynbos birds of prey

Because of their tendency to range widely, it is not easy to assign birds of prey strictly to particular fynbos vegetation types. Over 20 species occur regularly in the Fynbos Biome, but none are restricted to it. Those species which the casual visitor is most likely to see are described below.

The Rock Kestrel and the Blackshouldered Kite are widespread in southern Africa and elsewhere. Both sexes of the Rock Kestrel are deep rufous in colour, spotted black underneath. The male has a distinct grey head and tail, the female is less distinct. The kite is a soft, pale grey above, with white below. Conspicuous black shoulder patches are visible both in flight and when the bird is perched. If alarmed or threatened it wags its tail energetically. Both the kestrel and the kite hunt by hovering, a feature which immediately distin-guishes them from almost all other birds of prey. Small mammals are the favoured prey of both species.

The Black Eagle, an imposing jet-black species with a white 'V' on the upper back, is found predominantly in mountainous areas. Dassies comprise 90 per cent of its diet, which ought to make it a welcome predator on any farm. The call of the African Fish Eagle, albeit somewhat effeminate, evokes an 'image of Africa' for many people. That it is as likely to be encountered perched on an electricity pylon at a sewage farm as on a suitably rugged dead tree on the shore of a glistening, reed-fringed lake, does not seem to have dispelled the popular conception of this distinctive species. In the Fynbos Biome this eagle is confined largely to coastal areas or, inland, to farm dams.

A summer visitor from Eurasia to most of southern Africa, the Steppe Buzzard tops many a telegraph pole from which it drops onto its insect and small mammal prey. The Jackal Buzzard (named after its yelping call) is characteristic of Mountain Fynbos in particular. It is a handsome species, the adults having a broad chestnut breast-band, black head and distinctive black-and-white underwings.

Apparently benefiting from the increase in nest sites provided by the planting of alien trees, the Redbreasted Sparrowhawk occurs throughout fynbos. It is a dashing little hunter, taking prey (mainly birds up to the size of doves) on the ground or at speed in flight. Both sexes have rich rufous underparts and slate upperparts. The tail is boldly barred black and white.

Two large falcons are found in fynbos. The magnificent Peregrine is, by and large, a mountain bird, although it will occur wherever there are suitable nesting cliffs. It may even enter cities, including Cape Town and its suburbs, to hunt pigeons. These and other bird prey are caught in flight by the falcon, which may stoop at 380k/h in the hunt. A larger, more sedate species, the Lanner Falcon, is more typical of low-lying and agricultural land. Flocks of twenty or more birds may gather at feeding or watering sites.

Some other fynbos birds are now described according to the vegetation types with which they tend to be associated. It should be borne in mind that the majority of these birds, with the exception of the six endemic species, may be found in a variety of habitats elsewhere in southern Africa.

## Mountain Fynbos

Perhaps the most characteristic fynbos birds occur in Mountain Fynbos. The endemic Cape Rockjumper is a most handsome thrush-sized species, with a far-carrying *pee-pee-pee* . . . call, which is well worth tracking to obtain a view of the bird. The males have a bold white moustache over a black face and throat. The underparts are richly rufous, and the back is grey streaked with black. Although a bird of remote, rocky mountain tops, the rockjumper occurs conveniently close to the carpark atop Sir Lowry's Pass and on the roadside at Pringle Bay where Mountain Fynbos vegetation stretches down to the sea. Look for it among large scree slopes and rocky outcrops where it bounds over boulders in a spectacular fashion.

Two members of the canary family are endemic to fynbos. The Cape Siskin is a small, gregarious species which, although most common in Mountain Fynbos, also occurs in other fynbos vegetation types. If, at a distance it appears rather a dull bird, a closer look will reveal bright yellow underparts, and white spots on the tail and wing tips which are noticeable in flight.

A close view of a Protea Canary, on the other hand, will reveal it to be every bit as dull as it was at a distance. It does have a fairly distinctive pale throat and a double white wing-bar, however (an old name for the species is the Whitewinged Seedeater). It feeds alone or in small parties on *Protea* seeds and a variety of fruits and buds. The Protea Canary is an impressive mimic and will include in its broad repertoire the songs and calls of most of the other species which share its habitat (p. 93).

The thin, reedy jangle of the Orangebreasted Sunbird (p. 95) and the hoarse chatter of the Cape Sugarbird (p. 88) are amongst the most distinctive sounds of Mountain Fynbos. These nectar-feeding birds are found wherever their food plants (predominantly *Erica* species for sunbirds and *Protea* species for sugarbirds) are in flower.

Weighing only 10g (little more than a box of matches), the Orangebreasted Sunbird is an active little bird. It is a familiar sight darting between *Erica* bushes, probing the tubular blooms for nectar with its long, curved beak, and often perching conspicuously to advertise its presence to other sunbirds or to investigate any intruder, such as the surreptitious birder. The brilliant, iridescent plumage of the male sunbird adds to its appeal.

A rather sombre, brown bird, with an unexpected splash of bright yellow below its long tail, the Cape Sugarbird is a familiar occupant of any flowering Sugarbush, pincushion or *Mimetes* patch. The male's ridiculously lengthy tail (which can measure almost 400mm, five times its body length) is particularly conspicuous in the undulating flight-display. Three of the male sugarbird's main flight feathers are distinctly broadened into a bulge on the inner web. These cause a characteristic *frrrt-frrrt* sound to be produced when the wings are beaten rapidly in the display flight.

Two common Fynbos gamebirds are the Greywing Francolin and the Cape Francolin. Coveys of ten or more of the former are found in montane scrub, feeding on bulbs and insects. It is a small francolin, finely barred below with a broad, black-spotted collar. The Cape Francolin is a larger, darker species with red legs and a dark top-knot. It is particularly active and noisy (it has a crowing cackle) in the morning and evening, and can be very tame. The Kirstenbosch lawns are a favourite haunt of this engaging, if eccentric, bird.

Another distinctive bird, the Redwinged Starling, occurs widely in mountainous areas. It will visit Strandveld to feed on fruit, however, and has also taken to a life of scavenging from rubbish bins and carparks (where it will glean insects from car windscreens and radiator grills). The male is a glossy blue-black all over, but with a chestnut flash on the wings. The female is similar but with an ashy-grey head.

Not all Fynbos birds are as conspicuous as the Cape Francolin or Redwinged Starling. Finding and identifying the other species may be trickier, and a sound knowledge of bird calls and song is often more useful than a pair of binoculars. In the mountains a monotonous *tick-tick* call will perhaps be the only indication that Neddickys are present. This small warbler, pencil-grey below and brown above, is not as boring as its plumage would suggest and is worth enticing out of the undergrowth with a good *skish*. 'Skishing', making a sound like a tyre deflating staccato, is a recognised way of persuading the more uncooperative species to show themselves. It can prove embarrassing if other people are present, but can save hours of peering in vain into thick bushes (which can itself draw accusing looks or ribald comments from the uninitiated). Why this particular noise should interest the birds is not clear, but it certainly attracts many species of bushbird that would otherwise remain invisible.

In our experience, the most infuriating species in Mountain Fynbos is the endemic Victorin's Warbler. This small, golden-eyed, warm-russet-brown bird seems to have virtually abandoned all attempts at flight and spends its time hopping through the undergrowth, flitting its near-vestigial wings and singing loudly and not unpleasantly. Always, it seems, just out of sight. No amount of *skishing* or other lures (playing back its own song on a tape recorder, for example) will entice this bird into the open. For no readily apparent reason, however, and usually after hours of fruitless searching, one of its kind may suddenly leap (not fly) into the open and parade magnificently before its disbelieving audience. Nevertheless, the nearest most birders generally come to seeing this obstreperous creature is the quivering of successive branches from and to which the bird is hopping. For this reason, and not for want of trying, Victorin's Warbler is not illustrated in this book.

Found in similar habitat, and with similarly short wings to facilitate creeping through dense vegetation, the moustached and raggedy-tailed Grassbird (pp. 46 and 47), on the other hand, is much more amenable and will readily respond to a vigorous *skish*.

'After-hours' in Mountain Fynbos sees the emergence of nocturnal species. Although quite widespread here and elsewhere, many people only encounter a Spotted Eagle Owl (p. 5), sadly, as a road casualty. The much rarer and more powerfully built Cape Eagle Owl is the 'blocker' on many a birder's 'wanted' list. The locations of a number of traditional nest sites in western Cape Mountain Fynbos are jealously kept secret by conservationists. And rightly so. The owl illustrated on page 4 is destined to a life in captivity as a result of pellet-gun wounds – an ignominious fate for so noble a creature. The Fierynecked Nightjar is another species which is both common in fynbos and a regular traffic victim. Nevertheless, its familiar '*Good Lord, deliver us*' call may be heard even in suburban gardens against the incongruous background of the noisy city. Under such circumstances it is certainly an appropriate appeal.

## Kloof Woodland

Dissecting the slopes of much Mountain Fynbos are thickly wooded riverine kloofs. Their vegetation supports a variety of bird species which would otherwise be absent from these mountain areas. Sallying flycatchers, such as the Dusky and Paradise Flycatchers, perch on exposed branches awaiting flying prey, which they catch in mid-air, after a brief pursuit if necessary. Other insectivores, such as the Cape Batis (p. 71), hunt methodically by foraging through the foliage, often in the company of other woodland species including Barthroated Apalis and Cape White-eyes. Ground-foraging Olive Thrushes and Cape Robins rummage amongst the leaf-litter, while Cape Bulbuls guzzle fruit, such as those of Wild Peach in the canopy above. Incidentally, it is the shape of the leaf, not the fruit, that gives this tree its common name (p. 94). The Lesser Doublecollared Sunbird tends to feed more on insects and less on nectar than its congeners, and regularly searches for prey among the flowers of forest trees, such as the Butterspoon Tree.

Occasionally, less common species appear, such as the Lesser Honeyguide (it does not, actually, lead predators to bee nests, although other members of its family do), which we mistnetted at Swartboskloof (p. 71).

All these woodland birds may be considered fynbos species in the broadest sense, but all are dependent on a particular plant community within the biome. Many, notably Cape Robin and Olive Thrush, are also quite at home in well-wooded parks and gardens, having successfully adapted to life in alien trees and shrubs where the indigenous ones have long-since disappeared. The shrill contact-calls of a party of Cape White-eyes, although no competition to the traffic, are also a familiar sound along many a eucalypt- or *Hibiscus*-lined street in what may have been, many years ago, the haunt of sugarbirds and sunbirds.

## Renosterveld

So much Renosterveld has been turned over to agriculture that the natural bird community, comprising predominantly small insectivores, has been replaced by one of open-country species. Larks, pipits, plovers, Pied Starlings and a selection of granivorous canaries and sparrows feed among the furrows and wheatfields where scrubby vegetation and grasslands would once have supported a completely different suite of birds.

Redcapped and Thickbilled Larks occur in large numbers in freshly ploughed or harvested fields. The Clapper Lark occurs in indigenous and transformed vegetation. Its sombre plumage is redeemed to some extent by an entertaining display flight, which involves much wing-rattling and drawn-out whistling as it plummets earthwards.

The few remaining patches of natural vegetation are often characterised by 'LBJs' (Little Brown Jobs).

Dominant among these are two small insectivorous species – the Greybacked Cisticola and Spotted Prinia (p. 71). The virtually tailless Longbilled Crombec spends much of its time skulking in the undergrowth. The Karoo Robin is often seen only as a white-tipped tail disappearing into the bushes. It is quite excitable, however, perching on top of a bush or fencepost, calling loudly and flicking its distinctive tail. The Bokmakierie (p. 95) and Lesser Doublecollared Sunbird are two more brilliantly plumaged birds that may be found in any scrubby vegetation, even the tiniest Renosterveld relics.

## Strandveld

Yellow and Bully Canaries provide a welcome splash of colour in Strandveld, where the latter bird is commonly seen feeding on a variety of fruits and flower buds. Frugivorous species are commoner in Strandveld thickets than in any other vegetation type in the Fynbos Biome. Cape Bulbul and three species of mousebird may flock into fruiting Milkwood and other shrubs.

Our Strandveld study plot at Olifantsbos in the Cape of Good Hope Nature Reserve supported over ten times the number of individual birds found in a nearby Mountain Fynbos plot of the same size. The 36 species we recorded in the Strandveld plot included many not found in any other vegetation types in the reserve. In addition to the fruit crops, a major lure for the birds at Olifantsbos was a mass of Wild Dagga which, when in bloom, attracted Malachite Sunbirds in droves, with an estimated 550 there in December 1987. The seeds of this plant and of a number of grasses were much favoured by tiny Common Waxbills.

## Bird ringing

Sunbirds and waxbills were among the many species caught during our ringing operations at Olifantsbos and elsewhere. Bird ringing (or banding) is a widely used and valuable ornithological research technique involving the marking of birds with uniquely numbered metal rings. Subsequent reports of ringed birds help determine their movements and longevity. Although used primarily to study migration (European Swallows and shorebirds ringed in South Africa have been found in many parts of northern Eurasia, for example), ringing can also elucidate particular aspects of the biology of the more sedentary species, such as site and mate fidelity, territory size and lifespan.

Birds can be ringed as chicks (see the Fiscal Shrike on p. 80) or, more usually, as full-grown, free-flying birds. The latter are caught in very fine, almost invisible 'mistnets' strung between tall poles. Individual nets can be anything up to 18m long and 3m high. A flying bird hitting a mistnet (the term 'missed net' is also common parlance in ringing circles) creates a deep, soft pocket from which it is carefully extracted. Once safely 'in the hand' a bird is weighed and measured, moult (the extent of feather wear and replacement) is recorded and a ring is placed on one leg with special pliers (pp. 22 and 23). The bird's

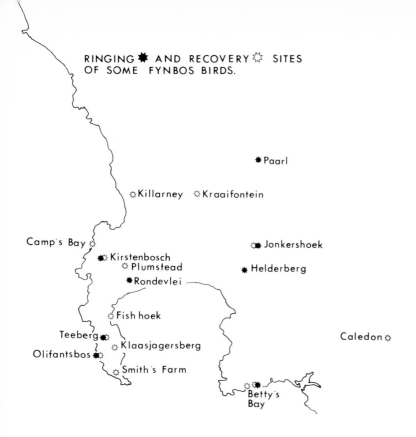

RINGING ✹ AND RECOVERY ☼ SITES
OF SOME FYNBOS BIRDS.

* Paarl

☼ Killarney   ☼ Kraaifontein

Camp's Bay ☼

☼✹ Jonkershoek

✹ Kirstenbosch
☼ Plumstead
✹ Rondevlei

* Helderberg

☼ Fish hoek

Teeberg ✹☼
Olifantsbos ✹☼

☼ Klaasjagersberg

Caledon ☼

☼ Smith's Farm

☼✹
Betty's
Bay

identity, age and sex are entered on a schedule, together with its unique ring number and details of where and when it was ringed.

In addition to the unique number, the lightweight, durable rings also carry a return address (SAFRING, UNIV CAPE TOWN or, on older rings, ZOO PRETORIA to which finders of ringed birds can report the ring number and circumstances (place, date and cause of death) of their discovery. By crosschecking with the ringer's records SAFRING (the South African Bird Ringing Unit) can determine the distance travelled and

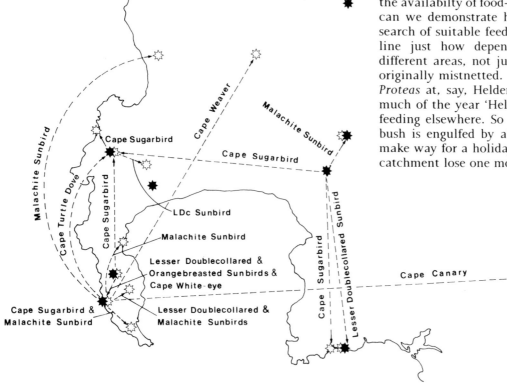

Malachite Sunbird
Cape Turtle Dove
Cape Sugarbird
Cape Weaver
Malachite Sunbird
Cape Sugarbird
Cape Sugarbird
LDc Sunbird
Malachite Sunbird
Lesser Doublecollared &
Orangebreasted Sunbirds &
Cape White-eye
Lesser Doublecollared &
Malachite Sunbirds
Cape Sugarbird &
Malachite Sunbird
Cape Sugarbird
Lesser Doublecollared Sunbird
Cape Canary

time elapsed since ringing. SAFRING then informs both finder and ringer of the bird's history.

To allow recognition of individual birds in the field, specific colour-ring combinations can be placed on one or both legs of birds in which the ringer has a particular interest (see p. 70). In our case, these are Cape Sugarbirds and sunbirds netted anywhere in the southwest Cape, or any species caught at our study sites at Olifantsbos and Swartboskloof. In this way, the origin of individual or groups of birds can be determined without the need to catch them again to read the ring number.

As the welfare of the birds is paramount, their extraction from mistnets (which can be very tricky) and 'processing' are carried out only by trained ringers who have served a long apprenticeship, or by trainees under the watchful eye of qualified personnel. Properly executed, however, the whole ringing operation has no ill-effects on the birds, apart from a few ruffled feathers which are soon preened back into place when the birds are released. The same cannot be said of the ringers, as anyone who has been punctured by the needle-sharp claws of a sugarbird, or gripped by the hooked beak of a tenacious Fiscal Shrike, will verify.

The success of any ringing project depends largely on the interest and help of members of the public who find and report ringed birds. The reporting rate of small birds is generally extremely low and, in the case of fynbos species, exacerbated by their often mountainous and restricted distributions. We have ringed over 5 000 birds in southwestern Cape fynbos and have been surprised by the relatively high number of recoveries (reports of ringed birds found away from the ringing site). The map illustrates the directions and distances of some of these. As our ringing efforts have concentrated on sugarbirds and sunbirds, it is not surprising that these feature prominently amongst our recoveries.

The nectarivores presumably migrate in response to the availabilty of food-plants, but only through ringing can we demonstrate how far each species travels in search of suitable feeding sites. Our recoveries underline just how dependent the birds are on many different areas, not just the one in which they were originally mistnetted. For this reason, to conserve the *Proteas* at, say, Helderberg alone is inadequate if for much of the year 'Helderberg' sugarbirds are, in fact, feeding elsewhere. So every time another pincushion bush is engulfed by alien vegetation or destroyed to make way for a holiday home, sugarbirds from a wide catchment lose one more feeding area.

# Fynbos reptiles and amphibians

Fynbos is home to a wide variety of snakes, lizards, tortoises, frogs and toads. These occur from the seashore to the mountain tops, in all vegetation types and in all habitats.

A warm spring afternoon is perhaps the best time to see reptiles. Tortoises are often encountered ambling along paths, and the observant fynbos visitor will notice that the little crests on many of the rocks are, in fact, sunbathing *Agama* lizards. The unobservant visitor, if particularly unlucky, may find himself planting his foot in a coil of belligerent Puff Adder. Rainy winter days will find fynbos chirping, clicking and croaking with frogs and toads, which are far less easy to see than to hear.

## Tortoises

Tortoises are comparatively well studied in relation to the other reptiles of the Fynbos Biome. They are generally absent from Mountain Fynbos, preferring Strandveld and Renosterveld vegetation. The rarest is the endemic Geometric Tortoise whose population has been reduced through habitat destruction to isolated remnants confined to a few of the remaining pockets of Renosterveld. This attractive tortoise (p. 107) is second only to a species found in Madagascar as the world's rarest and most endangered tortoise – not an enviable distinction.

The four or five other tortoise species in fynbos fare rather better, but none is confined to the biome. The commonest is the Angulate Tortoise which is found throughout much of the Cape Province. Like all tortoises it lays eggs, surprisingly large (up to 3cm in diameter) relative to the size of the parent (a big Angulate will measure less than 30cm in length), one or rarely two at a time in specially excavated burrows. Here the young develop for six to twelve months before hatching out. This must be a fairly laborious procedure, given the thickness of the heavily calcified eggshell.

Tortoise hatchlings are prey to a number of fynbos animals, including mass-attack by ants. Neither are adults immune to predators, and their hard carapace is no defence against the likes of the Black Eagle. We have seen an eagle clutching an Angulate Tortoise in its talons and drop it from a height of 30m onto a rock slab before gliding down to eat the shattered remains.

Unlikely to succumb in this way is the Leopard Tortoise which can weigh an impressive 40kg or more. This formidable species occurs naturally in the eastern parts of the biome, but has been widely introduced elsewhere. It may be encountered on many a fynbos nature reserve or wildflower garden path.

The Common Padloper, with its distinctive flattened shell and hooked upper jaw (an alternative name is the Parrotbeaked Tortoise), rarely exceeds 10cm in length. It is found virtually throughout the Fynbos Biome and north into the Karoo. The illustration on page 85 depicts one that we moved from the middle of a busy road and took the opportunity to draw before releasing it some distance into the veld. Although there can be little excuse for running over a tortoise (it's not as if they sprint out under the wheels of a vehicle) many are killed by careless or thoughtless drivers.

The only aquatic chelonian in fynbos is the Cape Terrapin. The title 'Cape' is a misnomer as, in common with a multitude of similarly named animals and plants, it occurs over much of southern Africa. Although it does spend most of its time in water, this terrapin is quite manoeuvrable on land. It is largely carnivorous (in contrast to the predominantly vegetarian terrestrial tortoises), feeding on fish, crabs and frogs and their tadpoles. The Cape Terrapin apparently makes fine eating if roasted whole, and the blood is imbibed as a reputed cure for 'fits'. Notwithstanding its culinary and medicinal properties, this terrapin remains fairly widespread in fynbos vleis. If these dry out it aestivates in a cool, underground chamber until the rains again flood the vlei.

## Lizards

So little is known about the ecology and taxonomy of fynbos lizards that it is even uncertain how many species occur here. Three species which we have often come across are illustrated on page 27. These are most frequently seen basking on rocks on warm days, but are very active and make a hasty retreat if approached too closely. One extreme escape mechanism was employed by the Orange-legged Skink which took refuge up my trouser leg. Happily, the two were none the worse after what was a novel experience for at least one of the parties concerned.

In the breeding season, the Rock Agama has a strikingly bright-blue head (p. 25), which it characteristically bobs when curious or being threatened. Its keen eyesight and rapid reflexes enable it to secure flying insects. The Three-striped Skink is also capable of impressive turns of speed, but tends to be more of a slinker, weaving slowly through the vegetation in search of food. It has taken to living in built-up areas and may be found in garden walls and buildings.

Also apparently quite at home in suburbia is the Cape Dwarf Chameleon. This most curious of reptiles can indeed alter its coloration to enhance its camouflage, but not to the extent that popular belief would have it. If provoked, the chameleon inflates itself and hisses defiantly. Coupled with the often garish body-patterning, this probably suffices to deter the majority of would-be predators. The ability to swivel the eyes independently is another disconcerting attribute, and one that enables the animal to keep an eye on all quarters. When an insect is targeted, both eyes are directed towards it. The rapid deployment of the sticky-tipped tongue is preceded by a certain amount of swaying and repositioning, presumably to fix the prey in its sights.

The chameleon may have a reputation for a number of things (including, according to some folklore, being deadly poisonous, which it is not), but a devoted parent is not one of them. The young illustrated on page 34 were simply dropped, still coated in a membranous envelope, by the mother and completely ignored thereafter. Within a minute or two of birth they had dried out and swaggered off purposefully to catch tiny flies and aphids for themselves.

The Knysna Dwarf Chameleon is endemic to the southern Cape and occurs in the transitional vegetation between forest and fynbos. Although often brilliantly coloured, some individuals, such as the one illustrated (p. 45), are less spectacular.

### Snakes

At least thirty species of snake are known to occur in the Fynbos Biome but none, apparently, is endemic. Contrary to popular belief, snakes are rarely encountered and, of those that are, only a few are potentially dangerous. This is not to say that the snakes are not there, rather that they are secretive, timorous, well camouflaged or retiring and have better things to do than terrify visitors.

As far as venomous species are concerned, in two years of fieldwork at the Cape of Good Hope Nature Reserve we saw Cape Cobra on perhaps twenty occasions and Puff Adder on less than ten. Some of these confrontations were, admittedly, a little too close for comfort, but the feeling was undoubtedly mutual. Mildly venomous or non-venomous species we saw there more frequently. With a single sighting of a Spotted (Rhombic) Skaapsteker, the year we spent in the Jonkershoek Valley must qualify as a herpetologist's purgatory.

Nevertheless, the danger *does* exist of being bitten by a poisonous snake and all reasonable precautions should be taken to minimise this risk. Those who romp barefoot through fynbos are asking for trouble.

Notwithstanding our own experiences, the Puff Adder is probably the most frequently encountered venomous snake in fynbos. It is widespread in southern Africa, and is responsible for the majority of fatalities involving stock and people. Slow-moving, volatile and generally bad-tempered, it relies on camouflage, not escape, as a form of defence, making it an easy target for the misplaced foot. The loud hiss (more resembling heavy breathing) that gives it its name, is not always forthcoming or may be concurrent with a strike. It is not easy to give the Puff Adder a good press, but it is certainly a beautiful, if rather sinister-looking snake. Generally less than one metre in length, the body is squat and the head broad and flat. The black-and-gold chevron patterning on the back makes it very difficult to spot in thick or sun-dappled vegetation.

An interesting adaptation of the harmless Common Eggeater is its apparent mimicry of the Rhombic Night Adder, a poisonous species. The intention is that would-be predators will leave the eggeater well alone, under the impression that it is not really an innocuous species. A further deterrent is a rapid coiling and uncoiling of its body which causes the roughly keeled dorsal scales to rub together, producing a suitably serpentine hissing. Because it feeds exclusively on eggs, the eggeater's teeth are greatly reduced in size and the cracking of the egg is effected by vertebral projections in the neck. So extraordinary is its feat of mandibular gymnastics that an eggeater the thickness of a finger can swallow a hen's egg. At Olifantsbos we regularly found Common Eggeaters on the upper beach and adjacent dunes, where a supply of Whitefronted Plover eggs was doubtless the attraction.

The Cape Cobra is a long (up to 2m) and graceful snake whose colour may vary from golden yellow through reddish-brown to very dark brown. We noted (from a distance) cobras of all three colour schemes on the Cape Peninsula. The cobra possesses highly poisonous neurotoxic venom which, in common with almost all snakes, it will administer only if cornered or unreasonably provoked. The rearing of the front third of the body and the classic display of the hood are somewhat awe-inspiring, but, given half a chance, the cobra will be off into the bush rather than confront whatever crosses its path. It preys on rodents, birds, frogs and lizards as well as other snakes.

A similarly large snake is the non-venomous Mole Snake, which may exceed 2m and is often unjustly persecuted in mistake for the Cape Cobra (although it has to be said that the persecution of *any* snake is unjust). The Mole Snake also occurs in a variety of plain colours, from sandy-brown to pitch black. Juveniles are usually light brown with pale-edged, dark brown-to-black spots or blotches. The Mole Snake is most common in Strandveld where it hunts molerats and golden moles, killing them by constriction. It is thus a very useful pest-control agent in agricultural areas.

The Boomslang is possibly less abundant in the Fynbos Biome than elsewhere in southern Africa, as it is almost exclusively a tree-dweller. The status of this, and many other, species is difficult to determine, however, because of its quiet and secretive habits. A back-fanged species with a potent haemotoxic venom, the Boomslang feeds on chameleons and birds and their nestlings. The colour and markings of Boomslangs vary considerably. Females are usually a muddy green-brown above, grubby white below. By contrast, the male Boomslang is a very beautiful creature, with bright green, black-tipped scales and a brilliant yellow belly.

The majority of fynbos snakes are perhaps less spectacular. The Southern Slugeater, for example, hardly conjures up an aura of glamour and mystery. Nevertheless, this innocuous little reptile is most endearing, in its modest way, and is perhaps more often seen (even to the extent of taking up residence in suburban gardens), than its larger and more charismatic brethren.

Whatever their appearance and lifestyle, all snakes deserve the greatest respect, both for their welfare and ours. Snakes are not the personification of evil that history and tradition would brand them. They are a fascinating and indispensable part of the environment and must be treated as such.

## Frogs and toads

There are approximately thirty species of frog and five toads in the Fynbos Biome. The most aquatic of the frogs are the platannas whose emergence onto land is restricted to occasional migrations. They are not the most attractive of frogs, but most of their features (streamlined, pear-shaped bodies, enormous feet and piggy eyes) are adaptations to their watery lifestyle. The Common Platanna, otherwise known as the African Clawed Frog (or Toad), was once known for its use in pregnancy testing: urine from a pregnant woman initiates, for a variety of curious biochemical reasons (outside the scope of this book), egg-production in the amphibian. This test has been largely superseded by artificial methods and the Common Platanna can now rest more easily in its fynbos pond.

Less assured of a future is the rare and endangered Cape Platanna (p. 39) which is confined to a few seepages and ponds on the southwestern Cape coast. One of its strongholds is the Cape of Good Hope Nature Reserve, but here this enigmatic frog is threatened by the introduction of alien fish which have, apparently, resulted in predation of its tadpoles. The spread of its near relative, the Common Platanna, has led to cannibalism and loss of genetic integrity through interbreeding. The action of waterside alien vegetation in altering water acidity also appears to have jeopardised the survival of Cape Platanna. As it is tolerant of high acidities in which the Common Platanna cannot survive, lowering of the acidity allows the latter species to colonise new ponds at the expense of the former.

A species with an even tinier distribution than the Cape Platanna is the Thumbed Ghost Frog of Table Mountain. Confined to only seven perennial streams which run through the wooded gorges on the mountain's steep slopes, it has the smallest distribution of any southern African amphibian. The head and body of this rather squat frog are flattened to enable it to secrete itself in narrow rocky crevices, and a cryptic patterning on the back also enhances its indetectability. Each finger and toe is equipped with a large pad or sucker enabling it to cling to slippery rocks in the fast-flowing water. The eggs of this species have never been discovered, but the tadpoles, which can take two years to metamorphose into frogs, have a flattened head and a large suctorial mouth to cope with life in the torrents.

Also a mountain-dweller, but not dependent on streams, is the Cape Mountain Rain Frog, one of at least five rain frogs found in the Fynbos Biome. As their name implies, these are most active at the onset of rain, at which time they emerge from their burrows, the males whistling shrilly to indicate their territory and to attract the females (who are, presumably, unaware that the source of their infatuation resembles an overblown potato with legs; p. 90). These globular creatures contradict many popular impressions of frogs; most notably they are completely averse to water, which they will never enter voluntarily. Not even the eggs are laid in water, but instead in an underground chamber where the complete metamorphosis from hatching through the tadpole phase to the froglet takes place entirely within the egg-capsule.

Confined to permanent waterbodies, the Cape River Frog's presence is indicated, more often than not, by a loud *plop* as it leaps from the bank into the water for sanctuary. The tadpoles of this species (p. 84) can grow to an enormous size (up to 13cm or more) and their development is delayed if food is scarce.

Our favourite fynbos frog is the Arum Lily Frog (pp. 74 and 75). It may be quite inconspicuous when positioned within the lily's chalice, which unsuspecting insects then visit, but brilliant orange webs, fingers and toes are displayed when the frog leaps. This species has a disjointed distribution from the southwestern corner of the biome to Mossel Bay, but is becoming increasingly rare through destruction of its wetland habitat.

Fynbos toads may be found some distance from water, although all require water for reproduction. The Raucous Toad is the most widespread toad in fynbos, appearing to be absent only from the Cape Peninsula. Seasonal puddles in sandy areas along the southwestern Cape coast provide suitable breeding habitat for the Cape Sand Toad.

Although many species of frogs and toads tend to be inconspicuous or nocturnal, all can be identified by their calls. We have found, however, that mystery species invariably become silent as soon as they are approached, rendering them permanently anonymous. Patience and perseverance are essential virtues of the amphibiophile.

## Insects and other invertebrates

An invertebrate is an animal without a backbone. Insects, spiders, scorpions and molluscs fall into this category. These humble creatures are, in all probability, hugely important contributors to the structure and functioning of fynbos, but have received scant attention in the past. The biology of many of even the most conspicuous and common species is still unknown, and a host of fynbos invertebrates are still undescribed taxonomically. It is difficult to assign a role to a particular creature if its affinities are unknown and if it hasn't even a name. Only butterflies, ants, bees and those species of economic importance (agricultural pests, for example) have been investigated in any detail. Even so, a researcher with a particular interest in oil-collecting bees was able to find twelve completely new species in the western Cape in less than two years. Should your life's ambition be to have an animal named after you, look no further than fynbos, where myriad insects and other creatures are waiting to be described.

Fynbos is rich in endemic species, many having affinities with ancient groups from other southern continents. Such creatures are considered to be a legacy of Gondwanaland. Table Mountain alone supports a number of endemic invertebrates including a cave cricket *Speleiacris tabulae*, a beetle *Colophon westwoodi* and an onycophoran *Peripatopsis alba*. The last of these is of particular interest, providing as it does an evolutionary link between legless worms and legged insects. Even more remarkable is a cave crustacean *Spelaeogryphus lepidops* which exists in one Table Mountain cave and nowhere else in the world. It is so unique that a special taxonomic order has had to be created for it as it fits into no existing one.

## Butterflies and moths

Of South Africa's total of almost 800 butterfly species, between 200 and 230 occur in fynbos, and it is one of the richest areas in the subcontinent for endemics, notably Satyrids (the 'Browns') and Lycaenids ('Blues' and 'Coppers'). Nevertheless, this is not, apparently, as great as expected, given the degree of plant richness and diversity. The fynbos butterfly fauna is largely made up of species that are also found in other regions. A quarter prefer savanna and bushveld, and so occur more commonly in the east where fynbos vegetation gives way to bushveld. Over 30 species favour grassy habitats (again, only found in the east of the biome), and about 20 are associated with evergreen forest patches and so become less abundant westwards. The remaining butterfly species are widespread elsewhere in southern Africa or the rest of the world.

The relative paucity of endemic species in fynbos (just over 20) is puzzling, but may be attributable to a tendency for butterflies to be more concerned with overall vegetation structure than plant-species composition. From the point of view of the flying patterns of the butterflies, many fynbos plant assemblages strongly resemble the temperate grasslands and Karoo whence many species have spread. The taller and more structurally complex fynbos vegetation types, with proteoid canopy, ericoid middle layer and restioid ground layer, seem to support very few butterflies. Less stratified, open vegetation supports far more. A shortage of larval food-plants (particularly the grasses on which many caterpillars feed) and the summer drought have also been held responsible for the low number of fynbos butterfly endemics. Again, as with the birds, the mobility of butterflies has not subjected them to local environmental pressures which have induced speciation of fynbos plants. On the other hand, those people who have themselves battled to make headway or even remain upright in a raging southeaster may have other ideas as to why there are so few butterflies in the southern Cape.

During our fynbos year the species we most commonly encountered was the Garden Acraea (p. 57) which we found everywhere from the streets of Cape Town to the high mountain slopes of Jonkershoek. In fact, this butterfly is found wherever its larval food-plant, the Wild Peach, occurs. The transparent forewings and orange hindwings (which vary in tone with

age) make this species and other members of its family easily recognisable. In common with many butterfly species, both caterpillar and adult are distasteful. Many would-be predators are deterred by the bright warning coloration, and the caterpillar has even resorted to chemical warfare, with its cyanide-charged hollow spines.

The Christmas Butterfly or Citrus Swallowtail is an unmistakeable species, boldly marked yellow-and-black with bright eyespots on the inner margin of the hindwing (p. 100). In the Jonkershoek Valley we regularly saw these swallowtails patrolling the forest tracks in such a way as to suggest they were defending a territory, as many butterfly species are known to do. Any such system appeared to disintegrate entirely when dozens of swallowtails were found feeding from the flowers of a Lucerne patch on disturbed ground. Many African Clouded Yellows were also found here, their caterpillars feeding on the foliage (p. 103). Because of its predilection for this fodder plant, this attractive species is one of the few agricultural-pest butterflies in the country.

A particularly well-studied community of fynbos butterflies is that of Table Mountain. Some 53 species occur here, which is nearly as many as are found in Britain. In all, 27 Lycaenids have been recorded, at least 12 of which display curiously enigmatic early life-histories. While feeding on their food-plants the caterpillars are attended by ants, for whom they produce a honey-like secretion on demand. Later, the caterpillars are transported to the ants' underground nests. To facilitate this operation the caterpillar either rolls up into a ball or locks jaws with the ant. Once underground, the caterpillars spare no efforts in taking advantage of their hosts' munificence, being not only treated as the ants' own offspring but also by eating the said offspring. This mildly unsociable behaviour appears to be tolerated by the ants because of a pacifying pheromone produced by the caterpillar.

Many of the world's Lycaenid butterflies exhibit this extraordinary relationship. In fynbos the underground stage (which may last up to a year) of their life-cycle serves to protect the caterpillars from fire, as well as from the universal array of surface predators and pathogens.

Despite sustained collecting in South Africa, new species of butterfly continue to be described each year. The majority of these come from the Cape, despite the fact that its natural history has been studied for much longer than the rest of the country. Simple observations and recording, of distribution and food-plants, for example, will add greatly to existing knowledge without having a detrimental effect on the butterflies. We cannot condone random and arbitrary collecting for its own sake.

The majority of moth species in fynbos, as elsewhere, are small, dull-coloured and difficult or impossible to identify by the uninitiated such as ourselves. In contrast to the generally sombre adults, the caterpillars of many species are magnificently large and garish. Those illustrated on page 75 do some justice to this claim, we trust. Hairs and bright colours serve to discourage predators (few birds appear to savour hairy caterpillars, the cuckoos being a notable exception), as do large false 'eye' spots. When disturbed, our Common Striped Hawk Moth caterpillars (p. 7) would rear up, retracting the head and displaying the eye-spot to best effect.

The Pine-tree Emperor Moth (p. 33) is one fynbos insect known to have benefited from the planting of pine trees. Its usual larval food-plant is the shrub *Rhus angustifolia*, from which it has switched enthusiastically to the totally unrelated pines. It has become a pest species in Cape plantations, where defoliation by the caterpillars can retard tree growth.

It has been suggested that there are more moth-pollinated flowers than there are butterfly-pollinated flowers in fynbos because nocturnal moths can avoid the fierce daytime winds, which tend to abate at night. The stronger and more direct flight of moths also gives them an advantage over the more cumbersome butterflies in this respect.

## Bees and wasps

The majority (probably more than 500) of bee species in South Africa do not conform to the popular impression of this insect. They are solitary, not social, and do not produce honey. Instead, they build small aggregations of nesting cells, often underground, of mud or paper made from masticated plants. A single egg is laid in each cell where, after hatching, the larva subsists on provisions of nectar and pollen deposited there by the female at egg-laying. The larva receives no subsequent parental care.

One fynbos bee which *is* social, the Cape race (variety) of the African Honeybee, also has idiosyncrasies that set it apart from other hive bees. The workers of this race lay eggs which can develop into other workers or even fertile queens, a task more usually accomplished by a single queen. In other hive-bee species, eggs layed by workers turn only into infertile drones. This curious Cape phenomenon has been interpreted as a response to heavy losses of queen bees on their mating flights, due to the strong summer winds.

The Cape Honey Bee is also notable for its docile nature, which endears it to bee-keepers. The African Honey Bee, in contrast, is a notoriously belligerent and bad-tempered beast. The Cape race is, therefore, of considerable importance as a reservoir of mild-mannered bees which can be crossed with races that are otherwise better adapted to conditions elsewhere but are unmanageable because of their irascibility. Some remote fynbos sites have been set aside as 'bee sanctuaries' to ensure that the Cape stock remains pure. The introduction of more aggressive or otherwise alien strains of honey bee to these areas is not allowed.

Carpenter bees are a familiar sight and sound in and around kloof woodland, particularly on warm days. These large, bumbling bees seek out Fountain Bush (p. 20), *Virgilia* and other leguminous-tree flowers from which to feed. One carpenter bee, *Xylocopa caffra*, is common in the Fynbos Biome. The female is black with two yellow stripes and about 20mm long. Males are the same size but completely covered in teddy-bear-like yellow fur. The female carpenter bee has a curious relationship with a tiny Dinogamasid mite, small numbers of which occupy a lentil-sized, chitin-lined chamber in the bee's abdomen. When the bee lays an egg in her specially hollowed-out nest chamber in dead wood, a few mites disembark. These lay their eggs on the food-mass the bee has laid up for her own young. On hatching, the mite and bee larvae share the food, and when the bees emerge from the pupal stage the mites secrete themselves in the pockets of the female bees. Males do not have these pockets and do not carry any mites. No satisfactory explanation is forthcoming for this intriguing commensalism. How many other similarly bizarre relationships are waiting to be uncovered in fynbos is anybody's guess.

Fynbos wasps may be solitary or social. Many build tiny nests of mud or paper in which to lay their eggs. Mud wasps such as *Sceliphron spirifex* lay up a stock of paralysed spiders with their single egg in a mud chamber, before sealing it up. The social wasps construct a delicate paper nest under an overhanging rock or bank (p. 60). Several females may combine to construct one nest, often containing 50 cells or more. The developing larvae are fed chewed caterpillar by the female and her daughters from previous nests.

Other wasps are parasitic and use their long ovipositors to 'inject' their eggs into live caterpillars and even the grubs of other wasps. Here their larvae develop, feeding on the innards of the living, and doubtless unwilling, host.

Fynbos wasps range in size from diminutive parasitic species less than a millimetre in length to the 35mm-long hunter-killer Pompilids. A female of the latter seeks out the likes of a Rain Spider (p. 14), paralyses it with a sting (which, one of us can confirm, is just on the bearable side of excrutiating) and then manhandles it backwards to her nest hole. Here a single egg is laid on the hapless spider, which is inexorably devoured by the developing wasp larva over the next week or so.

## Other creepy-crawlies

Ants are probably the most numerous above-ground invertebrates in fynbos. In all, 45 species have been recorded from the Jonkershoek Valley alone, with 32 species found at a single site. Here, the most abundant species was the Pugnacious Ant. The role of ants in transporting fynbos seeds is described later. They apparently are poor pollinators of flowers, but tend rather to thieve nectar.

Beetles make up the largest order in the animal kingdom. Some 300 000 species have been recognised worldwide and there are doubtless many still to be discovered, not least in fynbos, which supports a long list of endemic species. Many are considered to comprise the last relict of a temperate Antarctic beetle fauna that existed before the continents separated 140

137

million years ago. Beetles come in all shapes and sizes and fill every fynbos niche. They live underground, in dead wood, flowers, foliage or in water. Between them they eat anything and everything, from nectar to dung. Some, like the tiger beetle (p. 64) and water beetle, are predatory, eating other insects. The water beetle (p. 39) is so tied to an aquatic life that its legs are virtually useless for walking but serve as paddles instead. This beetle is a strong flier and travels between bodies of water under cover of darkness. The blister beetle contains the poison cantharadin, a property the beetle advertises to potential predators through its striking coloration (p. 18). This gesture is not entirely altruistic, however, as any predator (bird or lizard, for example) would have to swallow or at least chew the beetle before appreciating its distasteful qualities.

Extensive speciation has also been promoted in fynbos flies, paralleling the diversity of the plants on which many depend for food. The biome is particularly rich in species with elongated proboscids adapted for feeding from long-tubed flowers. The Tabanid on page 101 is such a fly.

Spiders are, quite unjustly, apt to fill many people with horror. Such people would not enjoy fynbos. Here there is a rich spider fauna, and an early morning walk along a mountain track can turn into a veritable 'hurdles', with a web to be negotiated every few metres. Certainly, many of these constructions are works of art too beautiful to sweep aside, that of the striking orb-web spider (p. 32) being an example. This spider's webs can be suspended between bushes two or more metres apart, and always have a distinctive zig-zag seam in the centre.

Less ambitious fly-traps are built by the silver vlei-spiders. The Rain Spider (also known as the Hunting or Ground Spider) actively hunts prey, rather than sitting in wait on or at a web. It is an accomplished construction engineer, however, amassing and binding large numbers of leaves into a nest the size of an albatross egg for its young (p. 14). Crab-spiders or flower-spiders come in a range of colours, depending on which particular flower they lurk under in wait for prey. When disturbed, these colourful animals make a hasty retreat by abseiling down to the sanctuary of the surrounding vegetation (pp. 50 and 69).

Spiders are a fascinating group and, like many fynbos animals with a grim reputation, certainly deserve a better public image.

Closely related to spiders are the scorpions. Being nocturnal, they are likely to be encountered only if you turn over boulders for whatever reason, thereby lifting the roofs off their houses. Many species are poisonous, but the most dangerous species are restricted to the most arid parts of South Africa and are not found in fynbos.

Some 60 species of snail and slug are endemic to fynbos, out of a total of 100 species – a very high proportion. A rich and distinctive leafhopper fauna also exists here, the high number of species resulting from the diversity of their food-plants. Topography and the tiny ranges of many of their host plants in turn limit the distribution of many of the leafhoppers.

Which leaves only dragonflies, damselflies, mantids, stick insects, grasshoppers, locusts, crickets, lacewings, antlions, cockroaches, ticks, fleas . . .

Rather than concentrate on specific taxonomic groups, some researchers have instead investigated the insects of particular plants. One such is the Sugarbush, which is of commercial as well as scientific interest. The plant can be divided into three niches: the flower heads or inflorescences that attract pollen- and nectar-feeders and seed-predators; the leaves, harbouring browsers; and the stems and seed-heads, borers. Of the 45 species of insect found in the flowers, 32 were beetles. One small species of beetle comprised 70 per cent of all insects found. Such beetles were discovered to be important pollen-vectors for the *Protea* – a role previously attributed only to birds, rodents and large beetles. The leaves were the subject of attack by 10 sap-suckers, 19 chewers and 3 miners. Ten different borer larvae were found in the inflorescence up to seed-head stage and 85 per cent of cone-stored seed was destroyed by insects within two years.

Such detailed studies are very much the exception rather than the rule. Other fynbos invertebrates must remain shrouded in mystery until such time as the scientific interest and resources are harnessed to investigate them thoroughly.

## Fynbos freshwater

Because of the porous nature of the soils, rain in upland areas of the Fynbos Biome tends to percolate quickly into streams which, in turn, drain into river valleys. Standing waters are, therefore, rare in the mountains, and are largely confined to spongy, peaty deposits and seeps that release water slowly. The flat coastal plains are similarly well drained and few permanent lakes exist there. Ephemeral streams and pans occur seasonally and are largely a feature of the western winter-rainfall areas. In addition to numerous small streams and tributaries, there are five major river systems in the Fynbos Biome. These are, from north-west to southeast, the Olifants, Berg, Breede, Gouritz and Gamtoos.

Although the streams which tumble down the rocky mountain slopes are clear, sweet and invigoratingly crisp, the water of many fynbos marshes, seepage zones and vleis is black and acidic. This is the result of high concentrations of humic acids and other toxic compounds, which are the breakdown products of the chemical herbivore-deterrents manufactured by many fynbos plants. These impart a deep, rich peaty colour and high acidity to many pans and stillwaters, so-called 'Blackwater Lakelets'. Those stillwaters that do not dry up in summer will at least warm considerably over this season. Many of the biome's waterbodies, like its soils, tend to be poor in nutrients.

The lives of fynbos freshwater animals are, therefore, fraught with extremes of temperature, food shortage, acidity and poisons, and they must contend with a water supply that may dry up in summer and flood or freeze in winter. The biome's waterbodies therefore support characteristic animal communities which have evolved to withstand these environmental pressures. Indeed, endemic species of insects with aquatic larvae comprise as much as 50 per cent of the invertebrates found in acidic western Cape streams. Less acidic waters appear to support fewer endemic species.

## Stillwaters

The few permanent stillwaters in the southwest corner of the Fynbos Biome are generally quite shallow (3m or less deep) and range from clear and alkaline to dark and acidic. The character of these vleis is largely determined by the surrounding fynbos vegetation and soils. Phytoplankton (microscopic floating plants) are abundant in the clear waters but rare in the dark ones, where the lack of light presumably limits photosynthesis. The low productivity of the Blackwater Lakelets is also reflected in the low numbers of larger animals utilising them.

At the Cape of Good Hope Nature Reserve, for example, although the Mountain Fynbos ponds contain a rare and interesting amphibian (the Cape Platanna), they support no indigenous fish, few invertebrates and even fewer larger animals (Cape Terrapins and not much else). They are also very infrequently visited by those species or numbers of waterbirds traditionally associated with shallow lakes. The odd Egyptian Goose (p. 31) or Spurwinged Goose are perhaps the most common visitors. In striking contrast, the alkaline, clearer waters of Rondevlei on the Cape Flats Strandveld support many ducks, geese, cormorants, pelicans, herons and egrets, as well as a variety of reed-bed and wetland warblers, cisticolas and other small birds in the marginal vegetation. All these birds are dependent on the abundance and availability of their food (water-plants and animals), which are, in turn, limited by seasonally variable levels of organic matter and water nutrients.

One of the most interesting and important waterbodies in the Fynbos Biome is De Hoop Vlei, a large brackish lake (620ha when full) near Bredasdorp. Here salinity and alkalinity fluctuate relative to season and water-inflow. As there is no surface outflow, water loss occurs only through evaporation and underground seepage. When the water level is high the vlei attracts large numbers of waterfowl, including up to 25 000 Redknobbed Coot. When the water level drops in summer the exposed muddy margins are frequented by migrant wading birds, including many species from northern Eurasia, such as Curlew Sandpiper, Little Stint and Ruff. Following flooding of the reserve in 1960, De Hoop achieved ornithological fame as the first breeding site in South Africa of the Greater Flamingo. About 800 pairs nested, rearing some 350 young.

## Running waters

A typical fynbos stream, from its source in the mountains to its outflow at the sea, can be divided into three zones. In its upper reaches it is rocky and fast-flowing. In the valley bottom the gradient flattens, the stream slows and the bottom silts up with mud and accumulated organic matter. Nearer its mouth the water will flow even more sedately, meandering through coastal dunes until it finally spills out into the sea or forms a wide, shallow lagoon. Such transitions, from the mountain torrent to the muddy estuary, can take place over a very short distance, and seasonal changes in water flow and volume can be immense.

The clear, cold waters of the mountains are very poor in nutrients, so aquatic animals rely on input from streamside vegetation for essential minerals. Decomposing leaves appear to provide most of the nutrients in this upper stream zone. Leaf-fall is greatest and stream flow at its lowest during summer in the southwestern Cape. Leaves tend to remain on the stream bottom for longer periods (although the average leaf pack has completely rotted away after 1,7 months) in this season, while in winter the leaves are soon washed away in spates before their nutrients are released. As a result, the number of aquatic insects and other invertebrates feeding on decomposing leaves in a study stream in the Jonkershoek Valley dropped from 105 per square metre in summer to only four in winter. Such animal communities comprise predominantly stonefly and blackfly larvae. Stonefly adults are fragile in appearance, fly rather feebly and are very short-lived (a few days at most). The immature nymphs live only in well-oxygenated, fast-flowing water where they may spend two years. Blackflies are small, blood-sucking insects whose larvae attach themselves to underwater rocks and sweep particles of organic matter into their mouths with long, fringed brushes. Mayfly larvae also occur in these upper reaches. Larger animals dependent on the decaying leaves include Freshwater Crabs which glean micro-organisms off the leaf surfaces (p. 21). The crabs, in turn, are eaten by mongooses, otters and the Giant Kingfisher. Few birds favour fast-flowing fynbos streams, but the African Black Duck may be found there, its young feeding on aquatic insect larvae.

Downstream, the abundance of aquatic invertebrates generally increases as more plant debris accumulates and the stony river-bed gives way to a muddy or sandy one. At this stage and farther downstream, however, there are virtually no rivers remaining in the Fynbos Biome that have not been radically altered by man. Here, run-off contaminated by fertilisers, herbicides, pesticides, sewage and refuse have disrupted water chemistry. Water extraction for domestic, industrial and, in particular, agricultural use, as well as alien vegetation on the riverbanks, dredging, damming and diverting, has so modified stream flow that one can only speculate as to how the rivers once ran their course and what plants and animals they supported in their original state.

## Fynbos fish

About 30 species of freshwater fish, the majority cyprinids (yellowfish, barbs and labeos), occur naturally in the Fynbos Biome. Most of the biome's major river systems contain endemic species, with the Olifants River system, which drains the Cedarberg and enters the Atlantic at Doringbaai, alone supporting eight. This is more than any other river system south of the Zambezi. The eight include three relatively large species, one of which, the burnished-gold Clanwilliam Yellowfish, may weigh up to 10kg and is revered as a sporting species. Five smaller species are the two rock-catfish and three redfin minnow species. Redfins, such as the Berg River Redfin, are small (seldom more than 10cm long), generally slow-growing fish which occupy the upper, stony, fast-flowing reaches of the rivers. Here they feed on insects and their larvae, and lay their eggs in well-oxygenated gravel beds. Their fins are particularly brilliantly coloured in the breeding season.

Other fish species found in the Fynbos Biome are more widespread. The Cape Kurper (p. 111) occurs from Verlorenvlei on the west coast to the Swartkops at Port Elizabeth. The Cape Galaxias prefers running water to still water (although it is tolerant of a wide range of temperatures and may survive for ten hours out of water on damp moss), and is found in most of the rivers from the Olifants to the Keurbooms east of Plettenberg Bay. It is a curiously transparent fish, with the vertebrae, heart, gills and body cavity being quite visible.

Far and away the biggest fynbos fish are the eels. Four species occur in fynbos, but all are also widely distributed elsewhere in South Africa. The Madagascar and African Mottled Eels may measure 2m and weigh 20kg. The two other species are half this size or less.

A large number of alien fish species has been introduced into virtually every waterbody in the Fynbos Biome. Where these have not exterminated the indigenous species, pollution and development have. The current status of the fynbos-endemic species, particularly, is very precarious. Already a number of species, including a redfin minnow in the Eerste River at Stellenbosch, have become extinct because of man's activities.

# Fire in fynbos

Sailing past Mossel Bay in 1497, the Portuguese explorer Vasco da Gama saw so much smoke emanating from the coast that he christened it *Terra de Fume*. What he probably observed were pastoralists burning the vegetation to improve the grazing for their livestock. Following European settlement, the colonists also adopted burning as a management tool, but by 1687 stringent laws were introduced to protect property and vegetation from the dangers of indiscriminate burning. That the problem was taken seriously is evident from the punishment meted out to miscreants: 'severe scourging' for a first offence, the death penalty for a second!

In the late nineteenth century, veld-burning contin-

ued to be considered undesirable. A succession of botanists in the first half of this century also advocated the complete protection of fynbos vegetation from fire, believing that it impoverished and desiccated the soil, exacerbated winter floods and summer drought and destroyed plant species. A strict fire-free management regime was thus adopted in private and State land alike at this time. Not until fairly recently was the ecological justification for this policy questioned and the role of fire scientifically investigated.

Complete exclusion of fire was virtually impossible, of course, and anyone who has seen the smoke billowing up from the Cape mountains will appreciate that fire is still very much a part of the fynbos year. The publicity that accompanies these often spectacular fires, invariably labels them 'destructive' and 'devastating'. This is not necessarily always the case, as there is no doubt that fynbos vegetation *is* fire-prone and, if it can burn, sooner or later it will.

## Fire patterns

Five 'fire climate-zones' are now recognised in the Fynbos Biome but, generally speaking and not unexpectedly, the vegetation will burn most readily when it is dry – mainly during summer in the winter-rainfall western areas and at any time of year under suitable weather conditions (notably berg winds) in the south-east.

Patterns of fire frequency and intensity vary from one fynbos vegetation type to another. Kloof woodland grows on fire-protected sites, such as rocky outcrops, scree slopes and river banks, and burns very infrequently and only under extreme weather conditions. Fires are similarly rare in Strandveld because of low fuel levels and a high proportion of succulent plant species which burn only with difficulty, if at all. Renosterveld in its putatively natural state (*Themeda* grassland) could possibly have burnt every year. So small an extent of Renosterveld now remains that a fire-regime as such hardly exists within it. Mountain and Lowland Fynbos will burn when sufficient fuel (living and, particularly, dead plant material) has accumulated to sustain a fire – generally between four and six years. As dead fynbos plant material decomposes very slowly, up to 40 tonnes per hectare of fuel may build up as the vegetation ages. With such amounts of combustible material, it is not surprising that Fynbos vegetation very seldom exceeds forty years old and that fires occur more often in this than in any other vegetation type in the biome.

The season, intensity, duration and frequency of fires, together with the age of the vegetation, topography, and recent and prevailing weather conditions, combine to make every fynbos fire unique. The 'natural' frequency of fires in the biome would now be very difficult, if not impossible, to determine, not least because their occurrence in the past was as likely to be random as cyclical. Natural fires still do burn large areas of fynbos every year, however. Rockfalls initiated by baboons, buck and earthquakes spark off fires as they must have done for thousands of years. Lightning

crops are burnt regularly, and block-burning of larger areas is carried out to reduce the high fire hazard of old vegetation. Management block-burns, predominantly under the auspices of the Cape Provincial Administration, currently operate on a 12–15 year fire-cycle. This regime is not recognised by all fynbos ecologists as representative of the natural system, however. To burn or not to burn is not an easy decision, when, in addition to the welfare of the natural vegetation, logistics, safety and economic considerations have to be taken into account. The hazards of even theoretically 'controlled' fires were forcibly demonstrated even as this was being written. A firebreak burn on the tinder-dry vegetation (much of it alien), swept through 1 800ha of fynbos and destroyed ten houses.

The effects of fire are as numerous as they are complex, but what evidence is there that fynbos plants and animals are, in fact, adapted to periodic burning? Do they show any traits which justify present burning-programmes in Mountain Fynbos?

## Fire and flowers

Fynbos plants exhibit four fire-survival strategies: regeneration from underground storage organs; protection of dormant buds by thick, insulating bark; resprouting from woody rootstocks; and survival as seeds which germinate after fire.

The efficient insulative properties of the soil prevent even the hottest burn from raising the temperature a few centimetres below the surface by more than a degree or so. Geophytic plants thus tend to survive fire very well. In fact, for many species, burning is a prerequisite for sprouting and flowering. Spectacular displays of *Watsonias* in bloom are a familiar feature of post-fire Mountain Fynbos (pp. 86 and 87). Synchronous post-fire flowering also minimises seed predation: so many seeds are produced at once that 'predator satiation' ensures there are more than enough for mice and other granivores to eat, leaving a good number to germinate later.

Many other geophytic species flower most profusely after fire, producing fewer buds in subsequent years or even becoming completely dormant. *Cyrtanthus* fire lilies flower only once following fire and become latent again until the next fire, which may be fifty years later. *Aristea, Agapanthus* (p. 104) and *Wachendorfia* (p. 89) are all examples of fynbos plants whose underground bulbs, corms or tubers are able to survive fire and the destruction of their above-ground stems and leaves.

The rapid 'greening' of Swartboskloof following the 1987 burn there was due largely to the speedy resprouting of Bracken from extensive rootstocks. The rate at which the shoots uncurled through the blackened soil was quite impressive. Many of the kloof's shrubs, although scorched and charred, belied their lifeless appearance by soon sprouting vigorously from stem-buds. The corky bark of Waboom (p. 16) insulates such

*Smoke billows up from a prescribed autumn burn at Swartboskloof*

strikes occur at an annual rate of only 0,2 to 3,4 per km² within the Fynbos Biome, the lowest incidence of ground-strike density in South Africa. Nevertheless, about 25 per cent of the known sources of ignition of fynbos fires are attributable to this phenomenon.

Once started, fynbos fires will burn as long as fuel, weather conditions, topography and the authorities permit. Formerly, fires were likely to have spread over considerably greater areas than they do today. A fynbos fire last century is reputed to have swept from Swellendam to Port Elizabeth, a distance of over 500km. Present-day fires rarely extend over large areas, and burns over more than a few days are rare. In the recent past, large fires have occurred in the Du Toit's Kloof Mountains in 1971 (180km² burnt), Baviaanskloof State Forest in 1975 (187km²), Kouebokkeveld Mountains in 1976 (300km²) and at Villiersdorp in 1984 (370km²). The last mentioned burnt for eleven days.

A high proportion of present-day fynbos fires are 'prescribed burns' carried out as part of veld management. Firebreaks around timber plantations and other buds, and bushes which had lost all their leaves in the flames displayed fresh growth within a few weeks. The resprouting Pepperbush, having survived the fire as a woody rootstock, was immediately the victim of another attack, this time from caterpillars of the African Monarch Butterfly, which lost no time in browsing the emergent leaves (p. 37).

A number of plant species survive fire only as seeds. Over 100 members of the Proteaceae, and one *Erica* (*Erica sessiliflora*) are among the 300 or more fynbos species which are 'serotinous'; that is, they shed only some of their seeds after every flowering season, retaining the remainder on the plant in fire-resistant storage organs such as hard, woody cones. A fire that kills the parent plant, dries and splits these storage organs, and the seeds are released en masse onto the bare ground. Here, germination may be enhanced by the absence of shading and competition for nutrients which would occur in mature vegetation. However, if one fire occurs hard on the heels of another, before serotinous plants have matured and set seed (which requires 3–6 years), then such species may become extinct. As their seeds are generally short-lived and do not accumulate in the soil, those produced by the previous generation are unlikely to survive long enough to germinate after fires in quick succession. Increased fire frequency in historical times is considered to have extinguished, locally or totally, many serotinous species. In the unlikely event of a fire-interval of more than 40–65 years, serotinous plants will also die out as their canopy-stored seeds are unable to outlive their parents, which by this time have become senescent and died. Thus, long intervals between fires also cause reductions in these species.

Recent research has shown that, in addition to fire frequency, the season of burn may be critical to serotinous-plant survival. Spring burns result in low recruitment (the addition of new individuals to the population, in this case to replace adults killed by the fire), as seedlings are unable to withstand the following summer's drought. Mortality of seedlings is also higher on north-facing (towards the sun) than south-facing slopes following late-autumn or winter burns. If water shortage inhibits germination until the following autumn or winter, the intervening months find the seeds exposed to high summer soil-surface temperatures and predation by small mammals. Two spring burns in succession can reduce the density of some *Protea* species from 10 000 to 100 plants per hectare.

A second important fire-survival adaptation of some fynbos reseeding (as opposed to resprouting) plants is myrmecochory. Derived from the Greek *myrmex*, ant, and *chorein*, to wander, this term describes the fascinating phenomenon of seed-dispersal by ants. Some 1 200 plant species in the biome have seeds with oily, fleshy appendages or coats called elaiosomes. When the ripe seeds are shed and drop to the ground, the elaiosomes exude a chemical lure which attracts ants, such as the Pugnacious Ant (p. 111). These industrious insects carry the seeds to their underground nests and there eat the elaiosomes. The seeds them-selves are too hard for the ants' jaws to penetrate, and so remain undamaged in the nest chamber. In this long-term storage they are safe from fire and the activities of surface-foraging mice and birds. There is also evidence that the viability of these seeds is maintained by the fungicidal properties of the formic acid which the ants secrete. In addition, the nutrient-rich soil of the nest enhances germination, which is stimulated by fire, and survival of seedlings. Although this method does not result in the seeds being moved very far (17m seems to be the furthest a myrmecochorous seed has been transported), it does guarantee that seedlings will find themselves within the same area in which their parent plant grew, a sure sign that environmental conditions are suitable for survival and reproduction. The disruption by alien Argentine Ants of this remarkable insect–plant relationship is discussed later.

## Fire and mammals

Small mammals display successional trends in response to fynbos fire, with changes in species composition and density taking place as the vegetation ages. The rates of recolonisation and recovery of these animals are very much dependent on the speed with which vegetation re-establishes.

Studies involving the live-trapping and marking of mice in Mountain Fynbos near George in the southern Cape indicate that fewer than expected were killed or even injured (singed whiskers being the order of the day) during the fire, presumably because such small animals are able to seek refuge under boulders and in burrows. Some of the first species to colonise a burnt area are also amongst the first to leave a year or two later, the Grey Climbing Mouse being one example. Other species apparently sensitive to the age of the vegetation are the Namaqua Rockmouse and the quaintly named Spectacled Dormouse, which colonise two to three years after fire, but disappear after a further two years. Species that occur exclusively in the older (10–30 years) vegetation near George include the insectivorous Greater Musk Shrew and Brant's Climbing Mouse. By contrast, some species (such as Spiny Mouse and the Forest Shrew) are less choosey, and may be found in Fynbos vegetation of all ages.

The study at George represents findings from only one site. On-going post-fire studies at Swartboskloof indicate that the small mammals there do not necessarily follow the same successional patterns and other trends of those near George. It is difficult to generalise in fynbos!

Little seems to be known about the effects of fire on large mammals in the Fynbos Biome, not least because there are so few large mammals remaining. It may be safe to assume, however, that both browsing and grazing species of buck and antelope would be attracted to the proliferation of fresh growth following a burn. Bontebok and Red Hartebeest at the Cape of Good Hope Nature Reserve certainly seemed more numerous in recently burnt areas. Although such a response would be expected, any apparent imbalance in animal numbers between young and old vegetation may be due to their increased visibility in the absence of cover. However, in the Renosterveld vegetation of the Bontebok National Park, grazing pressure was most severe within the first year after burning, decreasing significantly thereafter.

## Fire and birds

The composition of bird communities and bird density in the various fynbos vegetation types are certainly influenced by veld-burning, although mortality during a fire is probably limited to eggs and unfledged young. Watching the passage of the prescribed burn at Swartboskloof, we were surprised to note just how little immediate reaction there was to the flames and smoke. Cape Sugarbirds were feeding on Waboom inflorescences within a couple of metres of the flame-front, and birds as small as Cape White-eyes and sunbirds did not hesitate to fly through the murky palls to reach feeding areas. Cape Robins and Southern Boubous called and sang as if nothing was amiss (the Nero syndrome). The whole affair seemed to be of only minor inconvenience. The Swartboskloof management fire was a 'cool' one and progressed relatively slowly. A hot, fast-moving fire, of the kind that probably occurs under natural conditions, would doubtless merit more positive avoidance tactics.

The destruction of food-plants by fire results in the disappearance of the nectarivorous species (sugarbirds and sunbirds) which comprise such a large part of the avifauna of mature vegetation. At the Cape of Good Hope Nature Reserve an area of old (more than 15 years) Mountain Fynbos held small numbers of Greybacked Cisticolas and Yellowrumped Widows all year, seasonally augmented by Cape Sugarbirds and, particularly, Orangebreasted Sunbirds. Following fire and the removal of much of the standing vegetation, a completely different suite of birds was found. This comprised opportunist species such as Crowned Plover (p. 43), which is a characteristic species of many post-fire habitats, Plainbacked Pipit, Orangethroated Longclaw and other ground-foraging species. Sunbirds became extremely rare, not surprisingly, and one of the commonest birds before the fire, the Greybacked Cisticola, disappeared altogether. Three years after the burn, the Crowned Plovers departed and some of the pre-fire species began to trickle back. But not until the *Proteas* and *Ericas* had matured and flowered would the sugarbirds and sunbirds return in large numbers.

## Fire and reptiles

The short-term effects of fire on fynbos reptiles are largely unknown. Certainly snakes and lizards are able to survive by sheltering under rocks. There is even some evidence that tortoises may 'head for the hills' where protection from the flames may be found in rocky outcrops. Nevertheless, large numbers of dead tortoises are a feature of many fynbos fires, the sanctuary of the shell proving no match for heat and smoke. If fire occurs at the 'right' time, however, tortoises may, like many plants, survive through their offspring. The eggs of the Geometric Tortoise, for example, are generally laid in spring and hatch in April or May. If a fire took place after hatching, adults and young would be killed, but an autumn burn in February or March (the most likely time for a fire under natural conditions in the tortoise's range) would not endanger the buried clutches of eggs. The emergent hatchlings would also benefit from the regenerating and relatively lush vegetation.

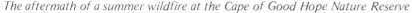

*The aftermath of a summer wildfire at the Cape of Good Hope Nature Reserve*

# Fynbos future

Future prospects for the Fynbos Biome, its plants and animals, are in the balance. The effects of human occupation, on the lowland areas in particular, have been so deleterious that major efforts are now required to save what remains of the natural vegetation and its inhabitants. Such efforts are rarely straightforward and invariably expensive.

To put fynbos conservation in perspective it is necessary to describe some of the changes brought about by man and the current problems facing the biome.

## Fynbos conservation in perspective

### Past and present persecution

In his diaries of 1652, Van Riebeeck recorded the presence of Black Rhinoceros on the slopes of Table Mountain and Hippopotamus in a swamp in what is now Church Square, Cape Town. Eland were common along the Liesbeek and Salt rivers and at Hout Bay on the Cape Peninsula. Buffalo occurred along the southern Cape coast as far west as Swellendam, and as recently as the last century Red Hartebeest were found throughout the Cape Province.

The first colonists also observed Elephants between Cape Town and Saldahna and in the Clanwilliam area. Such was the demand for ivory that by the eighteenth century the level of exploitation exceeded what the population could withstand. In 1736, one observer recorded that Elephants were being 'shot at continually', and in 1775 the explorer Sparrman reported hunters returning to Cape Town with their wagons laden with ivory from expeditions lasting eight or nine months. By this time the last of the Fynbos Biome's Elephants had taken refuge in the forests around Tsitsikamma. Despite attempts at protection and the pitifully low numbers remaining, persecution continued until the 1920s. In 1970, the last twelve Elephants of the fynbos region could be found in the Knysna Forest. Their pathetic fortunes are exemplified by the fact that now only three remain there. These represent some of the only truly 'wild' Elephants in South Africa, almost all others being confined to game parks.

The Lion is another species on the list of large mammals formerly occurring in fynbos but exterminated through persecution and man's appropriation of the land. Widespread in the eighteenth century, its numbers rapidly declined, and the last one was shot near Cape Town as recently as 1850. The Cape Lion was reputed to be a singularly large and dark race and it is interesting that many of the animals now in zoos around the world display these characteristics. It has been suggested that these are the descendants of Lions collected at the southern Cape last century and transported overseas. So the Cape Lion may live on, albeit somewhat ignobly.

One fynbos mammal that has been totally exterminated is the Blue Antelope. Originally occurring in a broad band from Worcester to Mossel Bay, by the eighteenth century it was present only in small numbers in the triangle bounded by Caledon, Swellendam and Bredasdorp, from which farmers with their firearms had no difficulty in exterminating it by about 1799. However, it appears that the Blue Antelope had travelled the path to extinction for some time before the arrival of Europeans. The introduction of sheep *circa* A.D. 40 and the consequent deterioration of the grassland component of fynbos vegetation, deprived the Blue Antelope of grazing and led to its decline.

About 18 stuffed specimens are all that remain of the once widespread and numerous Quagga. Its historical distribution, including its presence in fynbos, is confused by the inability of early travellers to distinguish it from the Cape Mountain Zebra. As one observer described zebras seen near Stellenbosch in 1685 as being 'of all colours with a wide blue stripe on the back and the rest of the body like that of a horse, with pretty, wide stripes of blue, yellow, green, black and white' we wonder how much credence to place on some early accounts! The Quagga was primarily an animal of the Karoo, and evidence that it did occur within the Fynbos Biome is equivocal. Whatever its previous status, it is now extinct through direct persecution and deterioration of the veld through overgrazing.

The Cape Mountain Zebra and the Bontebok escaped extinction by a hair's breadth. Historically, the Cape Mountain Zebra ranged throughout the mountains eastwards from Paarl. Human persecution then reduced their numbers such that by the 1930s the population numbered less than 50 animals. Government support for the establishment of a zebra reserve was unforthcoming, the Minister of Lands at the time being unwilling to help protect one of the rarest mammals in the world or, as he labelled them, 'a lot of donkeys in football jerseys'. Nevertheless, through the foresight and dedication of a small band of concerned farmers and conservationists a reserve was finally established at what is now the Mountain Zebra National Park at Cradock. The Cape Mountain Zebra population at present exceeds 400 individuals, and introductions have taken place to the fynbos reserves at De Hoop and, in 1986, the Cape of Good Hope.

The status of the Bontebok was similarly perilous. Again, persecution was responsible for taking the species to the edge of extinction. In a pioneering and enlightened conservation move well ahead of its time, the Van der Byl family set aside part of their farm at Bredasdorp in 1837 as a sanctuary for 27 Bontebok, an example then followed by adjacent landowners. The future of the species was thus assured, but only just. In 1931 the Bontebok National Park was established near Bredasdorp, but because of the nutrient-deficient pasture there the animals and the park (in name) were moved to their present location near Swellendam in 1961. Bontebok have now been relocated to other reserves and game farms. The entire population is at present still not much in excess of one thousand.

Competition with domestic stock and shooting for sport and food have caused many large animals to be exterminated in fynbos and virtually everywhere else in the South African subregion. Inevitably, land change and human population growth have made it impossible for these animals to re-establish naturally.

Is there any good news to be found amidst the doom and gloom surrounding many fynbos mammals? As far as the larger species are concerned the answer is not a lot.

Chacma Baboons are still quite widespread, but often come into conflict with man by raiding crops, plantations, houses and gardens. People have not helped the situation by feeding the baboons at the roadside. This practice leads the animals to associate people with handouts and they become aggressive if none are forthcoming. That the male (baboon) has canine teeth larger than a Leopard does not seem to deter the ardent baboon-feeder. Near Simon's Town we have watched in complete disbelief as parents thrust their crust-clutching children towards a hungry baboon. Such people not only endanger their children but jeopardise the future of the baboon. *Do Not Feed the Baboons* is a widely publicised and equally widely ignored slogan.

That arch-enemy of the baboon, the Leopard, is seen only rarely. However, an unusually tame individual at Betty's Bay in 1987 provided much excitement and pleasure until it was shot by a smallholder. In April 1988 the Directorate of Nature Conservation announced that an area of the Cedarberg had been set aside as a Leopard sanctuary where the animals are to be protected by law. This important move will, it is hoped, ensure the future of at least some Leopards in fynbos.

Conflict, real or perceived, with the stock-farmer is the undoing of many other carnivorous species. The poisoning and trapping of 'problem' animals also endanger non-target species, and many harmless animals, such as the Bateared Fox, are persecuted indiscriminately. It has been estimated that for every Blackbacked Jackal killed with poisoned bait, 110 non-target animals also die.

Birds, too, have suffered in this way. Anything with a hooked beak or talons was considered a threat to livestock. Eagles, buzzards, falcons and hawks were, and in many places still are, systematically shot, trapped and poisoned. The Cape Vulture now finds itself in a very precarious situation. Reduced to low numbers by direct persecution, the remaining few in fynbos currently suffer from a depressed food supply. This is a result of the extermination of game and the predators that would have left their prey carcasses for the vultures to scavenge. Present farming practices are unlikely to provide much dead stock for the vultures. Poisoning of stock carcasses (to kill predators or blowfly maggots) also kills scavenging vultures. In addition, because of the shortage of bone fragments discarded by mammalian predators, vulture nestlings receive insufficient calcium, and the birds are consequently weak or malformed. Collision with overhead wires and electrocution on power lines are further causes of mortality. Moves by the Electricity Supply Commission have contributed greatly to minimise these hazards through the redesigning of pylons and increased visibility of power lines. The only Cape Vulture colony remaining in the Fynbos Biome is at Potberg, east of Bredasdorp. Here the numbers of breeding birds have been declining throughout this century, and less than twenty pairs now remain.

While direct persecution is deleterious to specific animals and birds, destruction of natural habitat is a more serious threat to fynbos. Removal of the natural vegetation and transformation of its landscape have occurred on an enormous scale in the Fynbos Biome, and continue unabated. These largely irreversible processes deprive plants and animals of the unique environmental conditions which they need to survive. In most cases their requirements are too specialised to allow recolonisation of the man-modified habitat.

Opinions differ as to the extent of destruction of the natural vegetation. Figures ranging from 34 per cent to 61 per cent are proposed for the area lost through human activity – largely farming and urbanisation. The areas given in the following account come from a variety of often conflicting sources. Nevertheless, a great deal *has* been lost and the continuing rate of destruction is high and shows no sign of abating. The bold red areas on the map of the Fynbos Biome illustrate the scale of the problem, which is most severe on coastal lowlands, particularly Renosterveld.

As the majority of the land area of Fynbos Biome is in private hands, the survival of a multitude of plants and animals lies with individual landowners. Conservationists have reason to be grateful to the many farmers who not only leave aside portions of their property for wildlife, but actively protect and manage such areas for conservation. In practical terms, such action costs the landowner money, or, rather, reduces potential earnings from the farm, so a level of sacrifice is involved just to 'leave well alone'.

## Last of the lowlands

Although the Fynbos Biome occupies only 1 per cent of the area of southern Africa, it contains 69 per cent of the subcontinent's threatened plant species. Some 1 326 species are now classified as 'threatened', a term which encompasses status ranging from 'indeterminate' through 'vulnerable' to (ominously) 'recently extinct'. About 26 species fall into the last category.

The Cape Peninsula has the greatest concentration of threatened plants, largely as the result of 300 years of agriculture and urbanisation. On the other hand, Cape lowland areas support only a relatively small number of threatened plants. This does not reflect effective conservation in these areas; quite the opposite. As so much has been destroyed so quickly, botanists have had no time to record what was growing there before the vegetation disappeared. Because the rich shale soils are suitable for crops, mainly wheat, over 90 per cent of coastal Renosterveld vegetation has been completely destroyed. The boom in grain farming that began in the 1920s brought about a transformation of the landscape which, together with excessive use of fertilisers and ploughing of steep, erosion-prone slopes, has all but eliminated the indigenous plants and animals. Fragments of Renosterveld remain in isolation on rocky hilltops, but these, too, are threatened by overgrazing and agricultural advances which make it possible to cultivate all but the most precipitous slopes for vines and other crops. Such marginal areas could form important sanctuaries for indigenous species.

Financial incentives from the State are not forthcoming for farmers who conserve veld. In their absence, conservationists have urged that capital be made available for the purchase of land for reserves, particularly in the critically endangered lowland areas. Of the remaining coastal Renosterveld, 0,9 per cent is currently protected in nature reserves or mountain catchments. Provincial (public department) reserves account for a mere 8ha of protected west coast Renosterveld. Just 1,4 per cent of remaining sand plain Lowland Fynbos is currently protected, all of it within a single *private* nature reserve.

Strandveld does not fare much better. Although 40,2 per cent is undeveloped, only 6 400ha (1,0 per cent) is conserved. The remaining area is under threat, particularly from urban development. Western Cape lowlands overall support the most poorly conserved vegetation type in South Africa.

The Strandveld of the Cape Flats can lay claim to being the most maligned lowland system. Of a total of 381 rare plant species in the Cape Town region, 161 (42 per cent) are restricted to what is left of the Cape Flats natural vegetation. Despite its enormous conservation and scientific value, only about 250ha of the total area of 550km$^2$ covered by the Cape Flats are conserved, although even as reserves these areas are by no means secure from development.

## Mountain Fynbos conservation

Fortunately, the topography and poor soils of mountain areas militate against the type of intensive agriculture and development practised in the lowlands. The valleys are heavily utilised, however, and timber plantations (mainly of pine) now cloak large tracts of hillside. Although Mountain Fynbos is completely unsuited for grazing, stock are put out in it, especially in eastern areas. This leads to damage through overgrazing, trampling and subsequent erosion of steep slopes.

Wildflowers are an important commercial crop for many landowners in Mountain Fynbos. From the point of view of conservation, this should be a more acceptable form of exploitation. It has, however, been ruthless and irresponsible in many instances. Little attempt has been made to assess or reduce the damage caused by careless harvesting or pruning and trampling around the target plant. Present restrictions allow the removal of not more than 50 per cent of the blooms of an individual plant each year, and no picking is allowed for a year before a planned fire. Only if these rules are adhered to will sufficient seeds be available to establish the post-fire generation. Although it is in the farmer's interest to harvest within these levels, they are often greatly exceeded.

Some 70 per cent of Mountain Fynbos is privately owned. Landlords, therefore, have the potential to play a major role in its conservation and, indeed, many such areas are maintained as reserves. The bigger private reserves, such as Vogelgat near Hermanus and Elandsberg north of Paarl, include some of the most ecologically valuable and scenically beautiful areas of the Fynbos Biome. Individually and together, the scattering of smaller Mountain Fynbos sanctuaries is also indispensable. The Botanical Society is responsible for some real mountain jewels, and protected areas such as *Mont Fleur* are an example of the important contribution that every landowner can make to conservation.

Much of Mountain Fynbos is managed under the Mountain Catchment Areas Act of 1970 (which covers some privately owned land) and the Forest Act of 1984. These Acts combine to give protection to over 1 000 000ha of the 3 571 200ha of Mountain Fynbos. At 30 per cent, this makes it the best conserved vegetation type in South Africa.

The former Department of Forestry was at the forefront of conservation of these upland areas of the Fynbos Biome and was responsible for the management (including alien-vegetation clearance) of State-owned land held under the above Acts, and twelve nature reserves totalling 51 099ha. The 24 569ha of the Hottentots Holland Mountain Reserve make it the largest nature reserve in the biome. Four 'Wilderness Areas' have also been proclaimed, with nature conservation being the primary motivation. The largest of these is the Cedarberg (64 400ha). Responsibility for these areas now rests with the Department of Nature and Environmental Conservation of the Cape Provincial Administration.

Before you sit back comfortably, content in the knowledge that Mountain Fynbos, at least, is reasonably well conserved for your enjoyment, please read on.

Recent evidence indicates that what remains of the natural vegetation of the Fynbos Biome, the world's richest floral kingdom, will be totally replaced by alien plants within 100 years, if appropriate and adequate counter-measures are not taken now.

## What is an alien?

An 'alien', in ecological terms, is not from outer space but is a species of animal or plant that has been introduced *by man* to areas outside its natural range. The majority of aliens in fynbos have been imported from other countries, but some have been translocated from elsewhere in southern Africa. Once self-sustaining populations are established, some aliens spread rapidly in their new habitat. These aliens are termed 'invasive' and as such constitute a major threat to the survival of the natural occupants of the biome.

In 1652 Van Riebeeck requested his superiors in the Dutch East India Company to 'send us anything that will grow', which they promptly did. In the 300 years since then, the importation of foreign fauna and flora, by accident or design, has resulted in a transformation of the fynbos environment which has vastly exceeded that of the previous 700 000 years of man's occupation. Although fynbos is now more severely infested with aliens than any other South African biome, there is no evidence that it is actually more susceptible to invasion. The determining factor appears to be the greater length of time during which the Fynbos Biome has been exposed to invasion.

## Alien plants

Although fynbos is a predominantly treeless environment, quite extensive pockets of woodland once occupied rocky areas and river valleys. From the earliest days of European settlement at the Cape these were heavily exploited for timber. Van Riebeeck recorded with pleasure that on the slopes of Table Mountain grew 'forests with thousands of thick, fairly tall and straight trees' and that 'one could get thousands of complete masts for ships from them'. Within fifty years most of Table Mountain's forests had been felled. By the early eighteenth century, those of the Peninsula were all but destroyed and alien tree species were introduced from abroad to cater for the colonists' increasing demands for building-materials and fuel. Most of the plantations of Cluster Pine (from the Mediterranean region) on Table Mountain were established at this time. In fact, this species had been planted as early as 1680 at Franschhoek by the French Huguenots, to whom the scrubby fynbos must have looked rather dull in comparison to their well-wooded homeland.

A further impetus to the introduction of foreign woody plants was provided by the condition of the Cape Flats which, as early as 1702 (a mere 50 years after colonisation) had degenerated to driftsand as a result of overburning and overgrazing. Legislation was introduced in that year to prevent further destruction of the vegetation. In 1742 a gang of convicts and slaves was employed to 'tie down' the shifting sand by planting Wild Olive and indigenous grasses. These came to nothing, and it became necessary to erect a line of poles to indicate the route to travellers. Not until 1845 was the first hard road across the Flats completed, but within two years it was impassable. The Colonial Secretary, John Montagu, then imported Australian wattles and Australian Myrtle and had them planted in a long 'screen' to initiate dune formation.

In 1875 the Superintendent of Plantations, J. S. Lister, expanded the Cape Flats sand-stabilisation operation into a governmental forestry agency. Under his direction timber plantations, principally of Cluster Pine, were established. Port Jackson was planted to shelter the pines, while hedges of hakea were used to keep out animals. Port Jackson was also planted for its bark, in an attempt to develop a tannin industry. Many plantations were begun and the goverment awarded prizes for the largest areas planted. These were soon abandoned, however, in the face of competition from Natal, where the Black Wattle was proving more productive. This species did not grow well in the dry plantations of the southwest Cape but has become a major pest of riverine vegetation there. It has been wryly pointed out that while Natal rejoices in its bark, fynbos suffers from its bite.

The scale of planting alien trees within the Fynbos Biome was clearly enormous. For a variety of reasons, many of the plant species, which were imported and established with such enthusiasm, escaped from or were deliberately planted outside the confines of plantation, nursery and garden. Those that have proliferated now pose one of the greatest threats to the survival of fynbos plants and animals.

At the latest count, 109 species of terrestrial alien plant have been found in fynbos. These differ in the extent to which they have invaded the different vegetation types – Rooikrans (yet another Australian wattle) is most extensive in Fynbos and Strandveld, for example, with Longleaved Wattle prevalent along river courses. In the Albertinia district, Rooikrans took only 30 years to invade 41 000ha inland from the coast. Now, invasion by wattles is so extensive that infestations can even be discerned on satellite pictures of the Cape. Silky Hakea (p. 103) has infested about 4 800km$^2$ of Mountain Fynbos from Cape Town to George in 130 years; hakeas and pines together now infest over 7 500km$^2$ of Mountain Fynbos; 17 per cent of the remaining natural vegetation between the Berg River and False Bay is alien-infested; 64 per cent of 532km of southwestern Cape rivercourses are infested with Sesbania (a garden ornamental from South America); wattles and other thicket-forming aliens infest almost 9 000km$^2$ of Lowland Fynbos. The extent of invasion by herbaceous alien plants is largely unknown. The attractive but poisonous St. John's Wort (p. 102) has infested over 20 000ha of agricultural and disturbed ground in the Boland. One aquatic species, Parrot's Feather, a South American waterweed imported to decorate aquaria, is now widespread in western Cape wetlands.

What has made these alien plants so successful in the Fynbos Biome? How can they possibly take over a natural system which has been established for thousands of years?

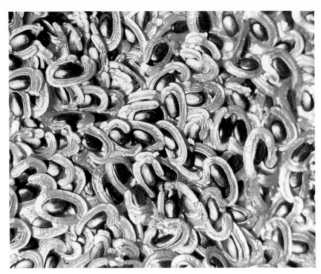

*Copious production of seeds has contributed to the success of many alien plants in fynbos. The nutritious appendage on the seeds of Rooikrans* Acacia cyclops *enhances wide dispersal by frugivorous birds.*

A great many alien woody plants have arrived in the Fynbos Biome from climatically and geologically similar environments, notably Australia, but also the Mediterranean region and South America. Their origin makes them pre-adapted to conditions in fynbos. Added to this are the absence from fynbos of the pests and diseases that act as controls in the aliens' native country, and the almost total inability of potential native predators and pathogens to utilise or infect the aliens. There are no shoot-wilting bugs, leaf-rotting mildews, stem-boring beetles, flower-munching caterpillars or seed-gnawing weevils to limit growth and reproduction.

Another weapon in the arsenal of many woody aliens is copious production of long-lived, heat-resistant seeds which are stored in the soil or released after burning when they germinate en masse. It has been estimated that up to 7,5 million winged seeds per hectare are released from a hakea stand following fire. The seeds of Black Wattle may survive for 50 years and 20 000 per m² may be found under a single tree. As many as 40 000 seeds per m² are found under some other wattle species. Higher spring growth-rates, aided by their ability to manufacture their own 'fertiliser' by assimilating atmospheric nitrogen, also give aliens the edge over native plants. Dispersal by birds and Chacma Baboons, which eat the fruits (as many as 19 000 per day in the case of the baboon) and later excrete the seeds, may also assist the spread of Rooikrans and other aliens in fynbos, where indigenous fruits are naturally rather scarce.

It has to be said that man has not only been responsible for the importation of the invaders, but has inadvertently aided and abetted their spread far beyond the points of introduction. The expansion of aliens in fynbos lowlands is enhanced by poor agricultural practices, leading to soil erosion and a prolifera-

tion of disturbed sites which encourage the establishment of alien plants. Transport of seeds in building-sand, disturbance by road and building development, and too-frequent burning of the indigenous vegetation all contribute to alien spread.

What, then, are the ecological effects of these invasive plants in fynbos? The repercussions of the alien inundation are many and varied: physical and biological, short and long term, subtle and blatant, and all of them dangerous.

Unable to withstand periodic flooding in the way that indigenous vegetation can, alien woody plants may accelerate soil erosion along waterways because flash floods rip out the aliens and expose the river banks to erosion. The rate of soil loss under pine plantations is higher than under indigenous vegetation. Alien woody plants may also reduce water flow from mountain catchment areas by means of rainfall interception and direct uptake from the soil. Stabilisation of natural dune plumes and sand-banks with alien plants has, in many instances, so influenced sand movements that the input of sand to beaches, vitally important in ecological and recreational terms, has completely stopped. Soil nutrient levels are modified by alien plants. As wattles can assimilate atmospheric nitrogen, the amounts of nitrogenous salts in the soil increase as a result. Phosphorus levels in Strandveld soils beneath aliens are some 50 per cent higher than under indigenous vegetation. Because fynbos plants are adapted to conditions where very low levels of nutrients pertain, this enrichment of soils has important implications for the survival or re-establishment of indigenous species.

Perhaps the greatest recognisable effect of the alien invaders is the massive drop in the number of indigenous plant species where they are physically swamped by the aliens. The sheer density of thickets and stands of aliens, with their deep leaf-litter blanketing the ground, makes it impossible for fynbos plants to survive in their shadow. As a typical hectare stand of 10-year-old Silky Hakea consists of 9 000 stems, three to four metres high, the absence of any other species is not altogether surprising. Parts of our study site at the Cape of Good Hope were so infested with Rooikrans as to be utterly impenetrable, and harboured not so much as an indigenous sprig in their gloomy interiors. In a 35-year-old pine plantation at Jonkershoek, indigenous plant cover was reduced by 55 per cent, plant density by 70 per cent and the number of species by 75 per cent from original levels.

The danger of aliens is possibly greatest to localised indigenous plant species. The restricted distributions of the latter already render them vulnerable to a variety of threats, of which alien infestation can be the proverbial last straw. Indeed, alien vegetation is the primary threat facing over half the endangered plant species in the Fynbos Biome.

As yet, relatively little assessment has been made of the effects of alien plants on indigenous animals, but some interesting findings have emerged from the few studies that have been carried out. Studies in dune fields infested with Rooikrans showed that only one species of rodent, the Striped Mouse, occurred. Five species were found on adjacent uninfested veld. However, on the Cape Flats the biomass of the Striped Mouse (the weight of all mice added together) was 5–10 times higher than in uninfested fynbos, the alien seeds obviously providing a richer and more abundant food supply.

From the point of view of many birds, alien vegetation can, dare we admit it, sometimes be considered 'beneficial'. All in all 36 species of bird have colonised the southwest corner of the Fynbos Biome and a further 27 have expanded their ranges within historic times as a result of the advent of alien vegetation. In an apparent response to the provision of nesting and roosting sites in alien trees, the Pied Barbet (p. 94) and Hadeda Ibis have recently moved rapidly westwards across the biome and are now happily ensconced as far west as the Cape Peninsula.

In general, however, alien thickets support fewer bird species than the indigenous fynbos vegetation, and many birds show a marked drop in density where the natural vegetation is replaced with alien monocultures. At the Cape of Good Hope Nature Reserve our bird censuses showed that the numbers of nectarivorous birds decreased significantly in thickets of Rooikrans. Here, the *Protea* and *Erica* food-plants of sugarbirds and sunbirds were completely eliminated by the alien. As these birds are potential pollinators of many fynbos plants, future prospects for this important bird–plant relationship are gloomy. Fewer birds and consequently low pollination levels of the surviving *Proteas*, for example, would result in low seed-set; this itself would exacerbate the decrease in indigenous plant density brought about by the alien infestation. The effects of alien infestation are often manifested in subtle, but nonetheless destructive, ways.

## Alien mammals

Few alien mammals have established in fynbos and of those that have, most – the farmyard and settlement dogs, cats, rats and mice – are commensal with man. The House Mouse reached South Africa from Europe in the seventeenth century. The Brown Rat was present, in Cape Town at least, as early as 1830. Both it and the House Rat are now well established at most ports. The European Wild Boar was released in southern Cape pine plantations in the 1920s to control the Pine-tree Emperor Moth, whose caterpillars feed on pine shoots before dropping from the trees to pupate underground. These pigs did not establish successfully, however, as it seems they faced competition from the indigenous Bushpig and were not immune to African swine fever. The control of caterpillars using pigs is an interesting early example of attempted biological control.

Cecil Rhodes was responsible for the introduction of a number of mammals including, via England, the American Grey Squirrel. This rodent has established only at places to which it has been translocated, and is presently restricted largely to the alien oaks and pines of parks and gardens from Cape Town to Swellendam.

The Grey Squirrel has been observed to chew the base of pincushion flowers and to eat the seeds of the indigenous Silver Tree. This may have serious implications for the future of the latter, a rare plant.

Any visitor to the Rhodes Memorial above the University of Cape Town campus will be familiar with the European Fallow Deer. Another of Rhodes's introductions, these animals breed freely within the confines of their park, which is maintained as such for their benefit and that of other alien mammals. It is not known how they or the indigenous vegetation would fare if they escaped into the adjacent fynbos.

One escapee, the Himalayan Tahr, *has* thrived. A pair of these Asian goats cleared the fence of their enclosure at the Groot Schuur Zoo in 1935. Finding the slopes of Table Mountain very much to their liking, they and their progeny flourished. By 1972 there were 330 tahrs on the mountain, causing considerable damage to the plant life through browsing, trampling and consequent erosion. Presently confined to Table Mountain, the tahr is unlikely to make its way, unaided, to the mountains beyond the Cape Flats. Translocations of other species to fynbos from elsewhere in southern Africa are generally restricted to the reintroduction of mammals which occurred in historical times but were extirpated by man, and the introduction of large mammals to nature reserves for public spectacle.

## Alien birds

A curious obsession of British and other colonists was to make any country in which they settled look as much as possible like the 'green and pleasant land' they had recently left. Cecil Rhodes was no exception, and enthusiastically set about 'improving' the amenities of his adopted Cape, which in the late nineteenth century had been branded as ornithologically uninteresting by the visiting English scientist G. E. Shelley. In 1897, Rhodes released European Starlings imported from England. How this noisy and pugnacious species was expected to 'improve' the Cape, goodness only knows, but it was Rhodes's most successful avian introduction, and now occurs throughout the southern Cape and as far north as Oranjemund on the west coast. As yet, there is no evidence that this bird is a threat to the larger, indigenous starling species, such as the Redwinged, although their feeding and nesting requirements are similar and may bring the two into conflict.

Rhodes's other introductions to his Cape Town estate of Groote Schuur included three European songsters – Blackbird, Song Thrush and Nightingale. None of these has persisted, although it was forty years before the Blackbird died out. Some 200 Rooks were also imported by Rhodes, doubtless for that 'English vicarage' atmosphere, but these soon succumbed. Marginally more successful was the Chaffinch, which can still be found in very small numbers along the eastern slopes of Table Mountain as far as Silvermine Reserve. Here it survives in pine plantations, a favoured habitat in its native Europe. Its numbers have never been high and it is certainly not invasive. A suggestion that the Chaffinch's range has been contracting lately is not supported by the individual we saw at the Cape of Good Hope Nature Reserve, some 30km south of the point of introduction, but it has taken a long time to get this far! To improve gamebird shooting, Rhodes translocated Helmeted Guineafowl to the western Cape. Here they have thrived, but agricultural development would have probably permitted the natural westward expansion of the species by now anyway.

The House Sparrow is another imported species which, although widespread in the biome, is pretty well restricted to urban areas and, in the country, to farms and homesteads. This sparrow arrived in Natal from western India as pets of sugar-cane workers in the late nineteenth century. After an initial period of slow colonisation in that province, it then exploded all over southern and south-central Africa. It was first recorded in the southwestern Cape in October 1962, at Piketberg, and reached Cape Town in December 1963. The House Sparrow seems to co-exist with the indigenous Cape Sparrow but may compete with the Cape Wagtail for food as both are, to some extent, scavengers.

Brought to the Cape by virtually the very first settlers was the familiar Feral Pigeon, which although now widespread in southern Africa, maintains its preference for city streets and buildings, and has not moved into untransformed areas of the Fynbos Biome to any great extent. It will, apparently, mate with the indigenous Rock Pigeon (pp. 66 and 67), if it cannot find a partner of its own species.

The European Mallard, a popular duck of wildfowl collections and now established on many southern Cape waterbodies, appears to hybridise freely with the indigenous Yellowbilled Duck. Such interbreeding has endangered the genetic integrity of native species elsewhere, notably New Zealand. Whether the Mallard is, or will become, a problem bird is the subject of some debate. The regal Mute Swan, a similarly popular and decorative species introduced from Europe, graced many Cape lakes and vleis until the early 1980s. A not implausible rumour explaining the birds' mysterious disappearance is that some imaginative entrepreneur trapped them and sold them to anyone prepared to pay R1 000 a pair. The subsequent depletion of the 'wild' birds put the price up to R2 000 until the species vanished!

In summary, the majority of the few species of alien birds that are established in the Fynbos Biome have a history of adaptation to the fields, streets, gardens and ponds of their Eurasian homelands. In fynbos they are similarly tied to man-modified areas. As such, there is no sound evidence that any alien bird has a detrimental effect on indigenous birds or other fynbos wildlife – as yet.

## Alien reptiles

There appears to be only one reptile in the Fynbos Biome that has been introduced from overseas. This is the Flowerpot Snake, so-called because it is thought to have arrived in pot-plant soil brought from Indonesia sometime before 1800, when that country and the Cape were under Dutch control. This creature still lives in Cape Town and its suburbs.

The translocation of reptiles from elsewhere in South Africa to fynbos is a more common phenomenon. Many southwest Cape nature reserves have populations of the lumbering Leopard Tortoise for example, brought by people for whom 'reserve' was synonymous with 'zoo' and, therefore, the perfect depository for unwanted pets and livestock.

Bibron's Gecko, a native of Little Namaqualand and Namibia, appears to be holding its own in the Cape Peninsula. About a hundred were released near Kommetjie some time before 1950. Further releases were made in the 1960s by a Fish Hoek pet-dealer who did not retain imperfect specimens.

## Alien fish

The history of the introduction of fish into the Fynbos Biome and southern Africa as a whole follows the sequence typical of so many Commonwealth or ex-Commonwealth countries: Goldfish for ornamentation, carp for food, and trout and bass for sport. In 1726, one traveller noted Goldfish in a tank in the Governor's house in Cape Town. By 1861, Common Carp had been introduced into the Eerste River, Stellenbosch, and the end of the century saw the arrival of Brown and Rainbow Trout, amongst other species, to benefit homesick colonial anglers for whom floating the dry fly was a much-missed pleasure. Bass have also been widely introduced for sport fishing. Bluegill Sunfish were liberated to provide food for the carnivorous bass and are now also found extensively in fynbos rivers. A number of indigenous fish species, such as the Cape Kurper, has also been translocated from one river system to another. These, too, may be considered alien.

Little is known about the harmful effects of introduced fish or other alien aquatic organisms in fynbos river systems. However, Brown Trout and Smallmouthed Bass are almost certainly responsible for the reduction and local extinction of seven endemic species of fynbos redfin minnows. The fish fauna of the Olifants River system is now completely dominated by bass, first released there in 1933. Only in the upper reaches of lesser tributaries, where waterfalls prevent the upstream movement of the invasive alien fish, can indigenous species be found in any numbers.

## Alien invertebrates

Through international trade, the arrival of invertebrate stowaways amongst cargoes and jettisoned ballast of ships from all over the world is inevitable. Numerous alien molluscs (including 17 species of snail in the Cape Peninsula alone) and the termite *Cryptotermes brevis* have arrived in the Cape this way. So too has the distinctive yellow-and-black-striped German Wasp, the attentions of which can make an ordeal of lunch at Kirstenbosch on a hot day. The potentially serious effects of this aggressive, communal-living species on the indigenous and generally solitary fynbos wasps and other insects have yet to be determined.

It is difficult to believe that one of the greatest long-term threats to fynbos comes from an ant. Not the innocuous creature it might appear, the Argentine Ant, through a combination of aggressive behaviour and feeding characteristics, poses a danger to the survival of fynbos plants perhaps second only to that of alien wattles, hakeas and pines. The Argentine Ant probably arrived at Cape Town in 1900, in a consignment of horse-fodder imported from South America by the British military during the Anglo–Boer War. It has since spread throughout the Cape and now even occurs in Johannesburg. Infestations by these tiny (2,5mm) brown ants are now common in households; here any domestic refuse or uncovered food soon swarms with them, and their interminable columns lead off to nests in parts of the building or garden. The ants are easily transported in rubbish and firewood, enabling them to spread and colonise virtually anywhere that man inadvertently carries them.

As early as 1915, myrmecologists detected a marked reduction in the number of indigenous ant species where the Argentine Ant was established. Of six dominant ant species found in one fynbos site without the Argentine, none was found at a site infested by the Argentine. This displacement of indigenous species by the more aggressive alien is at the root of the problem. As we have already mentioned, many fynbos plant species have their seeds dispersed by ants. These insects store the seeds in their underground nests, thus protecting them from fire and seed-predators. Argentine Ants, however, having removed the seed's fleshy elaiosome 'food packet', do not transport the seed underground, but leave it on the ground surface, where it may be burnt in the next fire if it has not already been eaten by mice or birds. The Argentine Ants thus compound the problem begun by their evicting the indigenous ants. The seeds of approximately 1 300 species of fynbos plant, including some of the rarest and most spectacular members of the Proteaceae, are dispersed and 'planted' by indigenous ants. Where these ants are excluded by the Argentine Ants, myrmecochorous Proteaceae showed total recruitment failure following fire. Such species will slowly, but surely, vanish as their seed-stores become depleted and there are no young plants to replace those that are old, senescent or destroyed by fire.

## Stemming the alien tide

What can be done to save fynbos plants and animals from the very real threats posed by this deluge of aliens? The early enthusiasm for bringing plants from other countries to the Cape and encouraging their spread has been matched only by their neglect when they invaded fynbos. Initial reaction to warnings of the threat they posed to fynbos was, at best, subdued.

Indeed, around the turn of the century the authorities seemed determined to achieve total cover of alien trees in the Cape Town area. Suspicions that alien plants were not entirely beneficial were voiced as early as 1888, when the government botanist Peter MacOwan branded Longleaved Wattle a 'pestilent weed', while the Conservator of Forests for the western Cape dismissed Port Jackson and Rooikrans as 'useless plants'. Dr Rudolf Marloth remarked, in 1928, that the authorities had missed all chances of containing the spread of alien plants, which were by then completely out of hand.

The first steps in tackling the problem of established aliens appear to have been taken by the private sector. As early as 1863, farmers in the Bathurst district met to discuss the spread of Silky Hakea on their land. A farmer at De Hoop initiated Rooikrans control in the 1940s when infestations were still relatively sparse. In the late 1950s a Committee for the Control of Alien Vegetation was established by the Botanical Society of Southern Africa to promote awareness of the problem and motivate for the clearance of key areas. The Mountain Club of South Africa, Cape Town Section (whose members, ironically, had previously spread alien seeds over many a mountain slope to provide future shelter and shade for climbers and hikers), was the first body to involve the public in organised alien-plant control work, through the formation of its 'Hack Group' in 1958. There are now twelve such groups whose industrious members strive to eliminate aliens from their local fynbos patches. Currently, many public and private bodies have assumed responsibility for alien clearance in particular areas.

The war against alien woody plants takes a variety of forms. Until recently these were largely mechanical, involving the chopping out of vegetation. For many aliens, however, elimination is not this simple. Species such as Port Jackson resprout vigorously from their stumps, necessitating herbicidal follow-up treatment. Burning is not a satisfactory method of alien clearance, as fire intensity can be increased dramatically by the additional and highly flammmable fuel load. The resultant hot fires can kill remaining indigenous plants and their seeds. Fire also stimulates the release and germination of alien seeds, resulting in a mass of seedlings much denser than the parent plants. This may again require herbicide application, a treatment often incompatible with conservation, but a necessary 'last resort' under the circumstances.

Different alien species may demand specific, integrated control methods, based on a knowledge of the plant's ecology and life-history. Hakea in mountain catchment areas, for example, is felled and left to dry for 9 to 12 months prior to burning. During this period the follicles split and the seeds are released, many to be consumed by rodents. Any seedlings that do sprout are then killed in the fire. Burning of the fynbos vegetation without prior felling of the hakea would only help it re-establish through fire-stimulated seed release. Too long a time between felling and burning would allow hakea seedlings to mature and produce their own seed.

*Clearing a dense infestation of wattles and eucalypts at Theefontein, Cape of Good Hope Nature Reserve. Few indigenous plants can survive in such thickets.*

Mechanical and chemical clearance methods are financially expensive, require heavy investment in equipment and manpower, and are a short-term or, at best, middle-term solution to the problem. Potentially harmful side-effects are produced by the use of chemicals and disturbance by men and machines. Nevertheless, pending the development of more efficient controls, such methods must be employed to stem at least the flow of invasion.

One such method, widely held to be the last chance for fynbos, is 'biological control'. This involves introducing the alien plants' natural enemies from their country of origin. Notable alien-attackers are insects which at some stage of their life-cycle feed on the flowers or seeds of the problem plant. The advantages of biological control are many. It is relatively cheap, only the target plant is affected, the control agents get on with the job themselves after initial introduction, and no chemicals or heavy machinery are involved that have the potential to inflict more damage than the aliens themselves.

Since 1962, the Department of Agricultural Technical Services, through its Plant Protection Research Institute, has been exhaustively researching biological control. Introduced insects are rigorously screened by the Institute and kept in quarantine until it is certain they will not attack any indigenous plants, before being released into fynbos. So far, biological control has been initiated on Longleaved Wattle, Silky Hakea, St. John's Wort, Sesbania and others, often with spectacular success. In the case of the wattle, a tiny Australian wasp *Trichilogaster acacialongifoliae* has been released which lays its eggs on the developing flower of the pest plant. The hatching grubs feed on the flower, which reacts by producing a fleshy gall (p. 76), in which the larva develops and pupates. Seed development and production are thereby severely curtailed and the future spread of the plant is restricted. As standing trees are not killed they can still serve their purpose as windbreak, sand-binder or source of firewood.

A three-pronged biological attack has been mounted against Silky Hakea, again using specially imported and thoroughly screened Australian insects. The larvae of the snout beetle *Erytenna consputa* prevent seed formation by feeding on the fruits of the hakea. When 20 of these beetles were released at one infested site in 1975, an impressive fruit loss of 81 per cent was recorded by 1981. Following fire in 1982 the density of hakea plants regenerating from seed was only 10 per cent that of the parent population. Normally after a fire the resultant stands of young plants are much denser than the parents. The larva of the hakea seed moth *Carposina autologa* complements the work done by the weevil, by attacking the seeds of mature fruits which have escaped the weevil's attentions. The hakea leaf weevil *Cydmaea binotata*, in turn, feeds on those seedlings that do sprout from any remaining seeds and severely hampers their growth and reproductive performance.

Such has been the success of these biological controls, in ecological and economic terms, that the hunt for more 'foreign agents' has been seen as a top priority in attempts to save fynbos vegetation. Laboratory trials undertaken with the seed-attacking beetle *Melanterius servulus* has proved it to be highly effective in the fight against spreading Stinkbean, a serious invader (from Australia) of indigenous kloof and

*Drastic measures are sometimes required to control alien plants in fynbos.*

stream-bank vegetation. As these biological controls are aimed specifically at the aliens' reproductive system, 'useful' weeds, such as timber species, can be safely cultivated without the risk of spreading into fynbos. Development of sterile strains and vegetative propagation could readily satisfy the need for new plants. Nevertheless, because *Melanterius servulus* may also attack the seeds of the Black Wattle, objections from the South African Wattle Growers' Union have caused further work to be suspended on the beetle or its release. Indeed, *all* research on every potential biocontrol agent of Australian wattles and related weeds was suspended in June 1987 because of the Board's objections.

There is a danger that biological control will be seen as an instant miracle cure, which it is not. However, the fact that it is highly selective, cost-effective, self-perpetuating and permanent, makes it a potentially powerful means of ending the long-term threat of the invasives. On this basis, the actions of the Wattle Board have serious implications for the future of fynbos.

Whatever control measures are adopted will need the full support of both public and State. Attitudes to aliens over the years have certainly changed. In the 1880s Table Mountain was described as having a 'bleak and naked appearance' by the Forestry Officer, who promptly planted it up with alien trees. As recently as 1936, the forestry authority pronounced that 'a large section of the public do not see, in a background of bare slopes and Gibraltar-like rock, the proper aesthetic setting for Cape Town and Table Bay'. How this 'section of the public' would react to the prominent and hideous tower blocks erected since can only be surmised. Today, the Cape Town City Council can be credited with devoting considerable money and manpower to clearing of alien trees from Table Mountain and replacing them with indigenous species. The process is a slow one and the transitional stages result in erosion, disfigurement and, indeed, an often 'bleak and naked appearance'. The operation must be viewed in the long term, however, with the reversion of the mountain to its natural fynbos-clad state the ultimate goal. Members of the public have opposed, however, the clearance of pines and eucalypts from the mountain's slopes, unaware that the aliens, majestic as they might be, destroy indigenous fynbos vegetation, create an added fire hazard, and reduce the flow and availability of ground and surface water. The rich fynbos plant communities are as important a heritage as the mountain itself.

The authorities at the Cape of Good Hope Nature Reserve appreciate that the reserve's primary importance lies in the conservation of its unique plants, and have also been extremely active in the control of aliens – largely Rooikrans and Port Jackson. Rondevlei Bird Sanctuary has seen a similar transformation in recent years, with dense Rooikrans and other wattles giving way to a rich assemblage of Strandveld and Lowland Fynbos plants, many of them highly endangered species. The Forestry Branch of the Department of Environment Affairs was, until recently, responsible for the control of large tracts of Mountain Fynbos catchment areas. Here it gave high priority to fynbos conservation and allocated funds each year to alien clearance. It is a curious fact that while one section of the department was fastidiously clearing aliens, another was actively propagating the very same weeds! The approximate numbers of seeds it sold in 1973 were: Port Jackson, 5,2 million; Rooikrans, 2,6 million; Australian Myrtle, 15,2 million. Only in 1975 did Forestry suspend sales of seeds of three wattle species and Australian Myrtle. Management, including alien clearance, of mountain catchments is now the responsibility of the Cape Department of Nature and Environmental Conservation.

The control of aliens in State land is made more problematic by re-infestation from adjacent private land where control is not carried out. The State has the power to remove alien weeds from private land up to 5km around any mountain catchment, but this still leaves a large area of fynbos where the eradication of aliens can be undertaken only if the landowner is able to afford it. Conservationists have proposed that, in the national interest, government should maintain private property alien-free at no cost to the owner. At present, it is estimated that only half of the fynbos infested with aliens is receiving any attention at all in terms of clearance efforts.

Cultivation of alien species that are already known to be invasive, and many more that are potentially so, will continue unabated until the public are made aware of the problem. It is certainly worrying that the 'virtues' of many alien plants are still expounded by nurseries, particularly when the benefits of the many beautiful indigenous plants suitable for cultivation have not been thoroughly investigated. We were horrified to hear recently a noted horticulturist singing the praises of a Mediterranean shrub, Oleander. Not only does this alien species form dense thickets that suppress fynbos riverine vegetation, but it is singularly poisonous. People have died after using a twig as a toothpick or pot-stirrer. A single chewed leaf can poison a child, and even honey from Oleander nectar and smoke from its firewood can be dangerous. If its invasive potential is not sufficient deterrent to its cultivation, then its hazardous health properties surely are. It is extraordinary that this shrub is still extensively planted in private gardens and along highways. Its cultivation in public areas provides another example of the apparent lack of communication between official departments, which sees some sections actively planting species that others are actively controlling!

South Africa is not the only country awash with aliens, of course. Many of this country's species have actually become invasive elsewhere in the world. Brother Berry, a shrub of Cape coastal dunes and sandy inland slopes, has become a serious pest in parts of Australia, for example. In Victoria it is a declared noxious weed where, ironically, it is choking out the indigenous Longleaved Wattle – the very same wattle that threatens a number of species here in South

Africa! A rare fynbos geophyte, *Gladiolus caryophyllaceus*, is now a pest in Australia. The Ice Plant, a widespread species in the southwest Cape, has become a serious weed in south Australian pastures. Up to 12 per cent of the dry weight of this curious plant is common salt (sodium chloride) and, when one dies, a miniature salt pan forms in which no other plants can grow until the salt is leached out by the rain. This may take many years in the dry climate of the region.

We must admit to finding it rather paradoxical that the dissemination of South African seeds throughout the world receives such enthusiastic support from not only the gardening but also the conservation fraternities. As fynbos is suffering chronically from the effects of imported ornamentals, should we not think twice before aiding and abetting potentially similar ecological disasters in other countries where the climate and conditions would generally favour fynbos plants?

The current status of alien mammals in fynbos does not require the exhaustive elimination measures demanded by alien plants. The one exception is the Himalayan Tahr. Culling of these feral animals on Table Mountain was begun in 1973, but because of the difficulty of the terrain and the expense involved, the numbers shot were lower than the growth rate of the tahr's population. Objections were voiced by people who considered culling cruel and unnecessary. If the animals are not eliminated, however, it is likely that the slopes of the Cape's most famous landmark will become even more disfigured than they are at present. Lack of public awareness and appreciation of its immense aesthetic and scientific value seems to be at the root of many of the problems affecting fynbos.

The stocking of nature reserves with alien large mammals should be easy to control but, nevertheless, still takes place. Animals are often introduced to cater for the assumed demands of the public, who are, however, presented with little information on what they should *really* be looking for in fynbos. In the past, herds of grazing animals, if they occurred at all, would linger only briefly in Mountain Fynbos, where grass is naturally very scarce. Confinement in reserves now precludes migration, and Mountain Fynbos plants suffer from the intense and persistent grazing of the animals, particularly after fire. Furthermore, the vegetation is low in nutrients, and animals obliged to feed exclusively on it frequently display symptoms of chronic nutrient deficiency – further evidence that the vegetation is unsuitable for stocking large mammals. The provision of fodder, such as Lucerne, only exacerbates the problem of alien plant invasion, and could introduce Argentine Ants and other alien invertebrates.

# What price fynbos?

Those people that invariably ask 'Why conserve?' are promptly regaled with the traditional offerings of 'untapped tourist and recreational potential . . . hundreds of hidden medicines and remedies locked up in its plants . . . great economic value of flowers and herbal products . . . indispensable supplier of clean water' and so on.

All these factors, as it happens, *are* pertinent to fynbos. But why do conservationists have to fight tooth and nail to save anything, particularly the most diverse, rich and beautiful gathering of plants in the world? It would be pleasant to believe that, inspired by the illustrations here, a walk through fynbos with a little time taken to search for the less demonstrative plants and animals would be enough to convince any sceptic that fynbos was worth looking after for its own sake.

However, as idealism generally produces moral support and goodwill but seldom any hard cash, we dutifully present those attributes of fynbos that make it *economically* worth saving. 'If it pays, it stays' is a depressing fact of life.

### The tourist trade

Despite being as far as it is possible to be from the gateway to South Africa (Jan Smuts airport), the Cape Peninsula is the country's top overseas tourist attraction. More accurately, Table Mountain is itself one of the most popular attractions, 300 000 people catching its cable-car each year. A further one million people use the mountain each year for easy strolls, more strenuous hiking or rock climbing. The Cape of Good Hope Nature Reserve also receives over 300 000 visitors annually.

Thousands of hikers make use of a network of trails established through fynbos mountains and along its coast. These may range from a brisk stroll around Fernkloof Nature Reserve, for example, to the four-or-five-day trails along the Tsitsikamma coast or the mountains of the Cedarberg. In 1985/6 a total of 65 597 'hiker nights' were spent on fynbos trails, and there is every indication that this figure is increasing.

It costs nothing to admire a view. Those visitors that come specifically to enjoy fynbos landscape, its plants and animals, do, however, spend a great deal of money on hotels, accommodation and other facilities.

The growing popularity of environmentally related leisure pursuits is confirmed by the increasing memberships of such voluntary conservation organisations as the Wildlife Society of Southern Africa, the Botanical Society of South Africa and the Southern African Ornithological Society. In addition, there are numerous smaller clubs and societies within the southern Cape that make direct use of fynbos for their activities, recreation and education. Many of these organisations include conservation as a main aim.

The proximity of as yet unspoiled fynbos to major centres of population, notably Cape Town, makes conservation of these areas even more important if they are to provide the 'breathing space' after which many urban-dwellers hanker.

### Flowers from the veld

Flowers are big business. Dried and fresh flowers

from fynbos are currently exported all over the world and the home market is blooming. *Proteas*, with their spectacular and unusual inflorescences and their long-lasting qualities, are particularly popular. And if any flower is synonymous with fynbos, then it has to be a *Protea*.

Direct harvesting from the veld is still carried out extensively, but for many species cultivation has now developed into an exact science. Though *Proteas* were propagated as long ago as 1774 (at Kew Gardens, in fact), demand for the species did not really take off until the 1960s. In 1965 the South African Protea Producers and Exporters Association (SAPPEX) was formed. Today SAPPEX has over 200 members who, in 1986 alone, exported over 2 000 tonnes of flowers, earning R7 million in foreign exchange. Ten new *Protea* cultivars were added to the country's export flower industry in April 1988. These displayed improved flowering properties, new flower shapes and stronger colours. The perfection of breeding and cultivation techniques not only makes propagation easier but takes pressure off rapidly dwindling and over-exploited wild stocks. Even so, the income from wisely managed harvesting can far exceed that from other land-use, such as grazing, of the same veld.

*Ericas*, *Brunias*, everlastings and restios are some of the many plants harvested for the cut-flower trade. The famous Chincherinchee is a fynbos ambassador that has decorated homes the world over since about 1880. Even in those days of sea travel the flowers survived the slow journey to Europe and were fresh enough to display on arrival.

Notwithstanding our reservations about the export of fynbos plants, hundreds of species of fynbos flowers provide bulbs and seed for the world's gardens. *Agapanthus* (p. 104), *Ixias* and *Freesias*, and a multitude of *Gladiolus* species are all fynbos in origin. *Nemesia* seeds were first received in London from the Cape in 1891, having been collected near Darling. Geraniums, the archetypal pot-plant, are fynbos in origin and must rank as some of the most popular of ornamentals. The list goes on, and there can hardly be a garden anywhere that does not sport a fynbos flower, whether the owners realise it or not.

Fynbos flowers *in situ* are also a valuable commercial asset. Thousands of visitors make the pilgrimage each spring to admire the spectacular wild-flower displays of west coast Strandveld and the adjacent interior. Specially presented wild-flower shows also attract visitors to many of the towns and villages in the area at this time of year.

## Other fynbos produce

As yet, few other products are harvested from fynbos. Rooibos tea is made from the dried leaves of *Aspalathus linearis*, formerly picked in the veld. 'Nature's great gift from the Cedarberg' is now cultivated commercially in and around Clanwilliam to satisfy increasing demand. Sales of Rooibos realised over R5 500 000 in 1986. Another fynbos special, made

from *Cyclopia* leaves is Honey Tea which, in common with Rooibos, contains none of the stimulants found in ordinary tea and coffee, but has soothing and relaxing properties. Even *rooineks* such as ourselves find Rooibos a very pleasant drink and deserving of wider appreciation.

The fragrant leaves of a number of fynbos plants contain oils which form the basis for perfume. Buchu *Agathosma* is used in home remedies, including Buchu brandy, for the treatment of various minor ailments. Whether the brandy or the *Agathosma* has the most therapeutic effect is debateable! Many more such fynbos plants must have medicinal or herbal properties that are waiting to be exploited – judiciously, please.

## Water

A reliable supply of good-quality water is essential for any urban or industrial area. Cape Town and Port Elizabeth are major industrial consumers and, with the recent petrochemical developments there, Mossel Bay soon will be.

Although the Fynbos Biome comprises only 4 per cent of the land area of South Africa, 19 per cent of the country's water catchments fall within its bounds. Maintenance of these catchments as providers of water heads the list of mangement objectives in Mountain Fynbos. Research has shown that water of a better quality emanates from areas where the indigenous vegetation persists. Replacement of fynbos vegetation by crops or invasive alien plants reduces water quality and yield. Up to 85 per cent of the annual rainfall in mountain catchments percolates out as run-off, and is therefore available to domestic and industrial consumers, compared to less than 10 per cent in non-catchment areas.

The availability of water is a major constraint on

*A reservoir and pine plantations in the Jonkershoek Valley*

development in South Africa. Reserving mountain areas as water sources is far more important than developing them. However, fresh, clean water has been the downfall of much Mountain Fynbos. The frighteningly speedy expansion of the population has so increased the demand for water and power that the valleys of the southwest Cape are being systematically dammed and flooded for reservoirs and pump-storage schemes.

## Research for conservation

Gone are the days of putting a fence round an area of veld and hoping that there is something worth conserving within it. Already many fynbos sites are recognised as requiring immediate protection, notably those that harbour the last individuals of a particular plant and those that support rich or vulnerable plant communities. Even assuming that government is prepared to provide the necessary funds, how can remaining fynbos and fynbos reserves best be managed for the benefit of their plant and animal occupants?

Fynbos ecology is nothing if not complex, and a misunderstanding of the specific requirements of a particular species may lead to its demise. What fire-regime to adopt is an invariably perplexing question. When populations of plants such as the Blushing Bride or Marsh Rose are reduced to a few individuals, management decisions could mean the difference between extinction and survival.

Although more questions remain unanswered, or

*Holiday homes at Cape Hangklip. Development and alien plants threaten to eliminate much of the southern Cape's natural coastal vegetation.*

*Indigenous vegetation has been replaced by agriculture over much of the Fynbos Biome's lowlands and mountain valleys.*

even unasked, than have been tackled, there is at present a very welcome growth in fynbos research. Within the CSIR's Fynbos Biome Project, for example, individual working-groups have been set up to investigate conservation of lowlands, palaeoecology, hydrology, and vegetation classification and mapping. Sites for intensive, multi-disciplinary research have also been identified. Swartboskloof has become the subject of feverish scientific scrutiny in over 30 projects. Climate and stream flow are continually monitored there, every plant seems to have a label, every bird a ring and every mammal a tag or radio collar! All in a good cause, naturally.

Long at the forefront of plant-conservation and horticultural research, the National Botanic Gardens at Kirstenbosch now boasts an endangered-plants research unit. This provides facilities for investigating methods of propagating critically threatened species when their future in the wild is in the balance. Seed-stores are maintained under controlled temperature and humidity regimes, and tissue culture allows cultivation even when seeds are not available. Such specialised techniques may be the only hope for many of the plants reduced to a few surviving individuals in the wild.

Many other bodies are deeply involved in fynbos research. University departments and institutes in the southwestern Cape take advantage of their setting for teaching and research, with fynbos as their laboratory. Students and fynbos conservation both benefit in this way. Herbaria co-ordinate plant-collecting and catalogue their often enormous collections of pressed flowers. These can include specimens collected two or three hundred years ago, which give valuable insight into former distribution. Botanical work is hampered by an acute shortage of taxonomists, with the result that some groups of plants have not been satisfactorily described systematically. As a result certain plant species could become extinct before they are even recognised!

In addition to formal research, many individuals and organisations make invaluable contributions to fynbos conservation. Hack groups, wildlife, bird and botanical clubs and societies, and school groups are all involved. Without them and without those individuals who have campaigned and lobbied for a better deal for fynbos for many years, the future of the biome would be considerably more bleak than it is now.

## Future prospects

In the long term, human population growth endangers every natural system in the world. Fynbos is no exception. The present population of South Africa is 28 million people and it is currently increasing at a rate of 60 000 per month. Projections of 47 million have been made for the year 2000. Twenty years on it will have reached 79 million. These people will all need land for housing, food, water, employment and leisure facilities. The pressures brought to bear on the natural environment for raw materials, living-space and re-creation, do not bear thinking of. Fynbos will not escape these pressures. A realistic population policy and commitment to conservation by government are thus urgently called for.

The future of fynbos now rests squarely with national and regional government. The expertise is available to advise on key problems, and areas and species have been identified that require immediate and far-reaching action to save them from disappearing altogether. It is up to the authorities to make sufficient funds available to harness this expertise, for the purchase and management of reserves, for the rigorous and effective enforcement of environmental legislation, and for the responsible education of the population in environmental matters.

What is happening to the Fynbos Biome now is exactly what has happened to many of the world's wild places in the past. The almost unparalleled opportunity therefore exists for South Africa to conserve an ancient and globally unique natural system. It remains to be seen if the authorities are prepared to take this opportunity. It will not come again.

Meanwhile, despite the often gloomy prospects, those stalwart individuals and organisations with a commitment to fynbos conservation press on. Please give them and 'The Smallest Kingdom' your support. Without it, *A Fynbos Year* will become a history book.

*Erica plukenetii*

# Field guides and reference books

## Plants

Baker, H.A. and Oliver, E.G.H. 1967. *Ericas in southern Africa.* Cape Town: Purnell

Bond, P. and Goldblatt, P. 1984. Plants of the Cape Flora: A descriptive catalogue. *Journal of South African Botany,* Supplementary Volume 13: 1–455

Burman, L., Bean, A. and Burman, J. 1985. *Hottentots Holland to Hermanus: South African wild flower guide 5.* Cape Town: Botanical Society of South Africa

Coates Palgrave, K. 1977. *Trees of southern Africa.* Cape Town: Struik

Jackson, W.P.U. 1980. *Wild flowers of the fairest Cape.* Cape Town: Timmins

Jackson, W.P.U. 1982. *Wild flowers of Table Mountain.* Cape Town: Timmins

Le Roux, A. and Schelpe, E.A.C.L.E. 1981. *Namaqualand and Clanwilliam: South African wild flower guide 1.* Cape Town: Botanical Society of South Africa

Levyns, M.R. 1966. *A guide to the flora of the Cape Peninsula.* Cape Town: Juta

Maytham Kidd, M. 1983. *Cape Peninsula: South African wild flower guide 3.* Cape Town: Botanical Society of South Africa

Moll, E.J. and Scott, L. 1981. *Trees and shrubs of the Cape Peninsula.* Cape Town: Eco-lab Trust Fund, Department of Botany, University of Cape Town

Moriarty, A. 1982. *Outeniqua, Tsitsikamma and the Eastern Little Karoo: South African wildflower guide 2.* Cape Town: Botanical Society of South Africa

Rourke, J.P. 1972. *The proteas of southern Africa.* Cape Town: Purnell

Vogts, M. 1982. *South Africa's Proteaceae, know them and grow them.* Cape Town: Struik

## Mammals

Cillie, B. 1987. *A field guide to the mammals of southern Africa.* Sandton: Fransden

Maberly, C.A. 1986. *Mammals of southern Africa: A popular field guide.* Craighall: Delta

Smithers, R.H.N. 1983. *The mammals of the southern African subregion.* Pretoria: University of Pretoria

Smithers, R.H.N. and Abbott, C. 1986. *Land mammals of southern Africa: A field guide.* Johannesburg: Macmillan

## Birds

Berruti, A. and Newman, M. 1986. *Discovering birds.* Cape Town: Struik

Berruti, A. and Sinclair, J.C. 1983. *Where to watch birds in southern Africa.* Cape Town: Struik

Cape Bird Club. 1981. *A guide to the birds of the southwestern Cape.* Cape Town: Cape Bird Club

Frandsen, J. 1982. *Birds of the southwestern Cape.* Sloane Park: Sable

MacLean, G.L. 1985. *Roberts' birds of southern Africa,* 5th edition. Cape Town: Trustees of the John Voelcker Bird Book Fund

Newman, K. 1983. *Birds of southern Africa.* Johannesburg: Macmillan

Sinclair, J.C. 1987. *Field guide to the birds of southern Africa.* Cape Town: Struik

Steyn, P. 1982. *Birds of prey of southern Africa.* Cape Town: David Philip

## Freshwater Fish

Bruton, M.N., Jackson, P.B.N., and Skelton, P.H. 1982. *Pocket guide to the freshwater fishes of southern Africa.* Cape Town: Centaur

Hamman, K.C.D. and Gaigher, C.M. *Freshwater and estuarine fish of the Cape Province.* Cape Town: Cape Department of Nature and Environmental Conservation

Jubb, R.A. 1967. *Freshwater fishes of southern Africa.* Cape Town: Balkema

## Reptiles and Amphibians

Boycott, R.C. and Borquin, O. 1987. *The South African tortoise book.* Cape Town: Southern Books

Broadley, D.G. 1983. *Fitzsimons' Snakes of southern Africa.* Craighall: Delta

Marais, J. 1985. *Snake versus man.* Johannesburg: Macmillan

Passmore, N.I. and Carruthers, V.C. 1979. *South African frogs.* Johannesburg: Witwatersrand University Press

Patterson, R. and Bannister, A. 1988. *Reptiles of southern Africa.* Cape Town: Struik

Rose, W. 1962. *The reptiles and amphibians of southern Africa.* Cape Town: Maskew Miller

Visser, J. 1979. *Common snakes of South Africa.* Cape Town: Purnell

## Insects

Claassen, A.J.M. and Dickson, C.G.C. 1980. *Butterflies of the Table Mountain range.* Cape Town: Struik

Dickson, C.G.C. and Kroon, D.M. 1978. *Pennington's Butterflies of South Africa.* Johannesburg: Donker

Germishuys, H. 1982. *Butterflies of southern Africa.* Johannesburg: Van Rensburg

Londt, J.G.H. 1984. *A beginners guide to the insects.* Durban: Natal Branch of the Wildlife Society of Southern Africa

Migdoll, I. 1987. *Field guide to the butterflies of southern Africa.* Cape Town: Struik

Pinhey, E.C.G. 1975. *Moths of southern Africa.* Cape Town: Tafelberg

Quickelberge, C. 1986. *Familiar South African butterflies.* Durban: Natal Branch of the Wildlife Society of Southern Africa

Scholtz, C.H. and Holm, E. 1986. *Insects of southern Africa* Durban: Butterworths

Skaife, S.H. 1961. *The study of ants.* Cape Town: Longmans

Skaife, S.H. 1979. *African insect life.* 2nd edition revised by J. Ledger and A. Bannister. Cape Town: Struik

## Alien Plants and Animals

Bruton, M.N. and Merron, S.V. 1985. *Alien and translocated aquatic animals in southern Africa: a general introduction, checklist and bibliography.* South African National Scientific Programmes Report 113: 1–71

Macdonald, I.A.W., Kruger, F.J. and Ferrar, A.A. (eds) 1986. *The ecology and management of biological invasions in southern Africa.* Cape Town: Oxford University Press

Stirton, C.H. (ed) 1978. *Plant invaders. Beautiful, but dangerous.* Cape Town: Department of Nature and Environmental Conservation of the Cape Provincial Administration

## General

Day, J., Siegfried, W.R., Louw, G.N. and Jarman, M.L. (eds) 1979. *Fynbos ecology: A preliminary synthesis.* South African National Scientific Programmes Report 40: 1–166

Deacon, H.J., Hendey, Q.B. and Lambrechts, J.J.N. (eds) 1983. *Fynbos palaeoecology: A preliminary synthesis.* South African National Scientific Programmes Report 75: 1–216

Kruger, F.J., Mitchell, D.T., and Jarvis, J.U.M. (eds) 1983. *Mediterranean-type ecosystems: The role of nutrients.* Berlin: Springer-Verlag

Lighton, C. 1960. *Cape Floral Kingdom.* Cape Town: Juta

Moll, G. 1987. *Table Mountain – A natural wonder.* Kirstenhof, Cape Town: Western Cape Branch of the Wildlife Society of Southern Africa

Van Rensburg, T.F.J. 1987. An introduction to fynbos. *Bulletin of the Department of Environment Affairs* 61: 1–56

Werger, M.J.A. (ed) 1978. *The biogeography and ecology of southern Africa.* The Hague: Junk

# Index of species

Scientific species names appear in *italics*, Afrikaans names in brackets. *Italic* numerals refer to illustrations.

## Plants

## Mammals

Clanwilliam Yellowfish  *Barbus capensis*  (Clanwilliam-geelvis)  140
Madagascar Mottled Eel  *Anguilla marmorata*  (Madagaskar-bontpaling)  140
Redfin minnow, Cyprinidae  140, 150
Rock-catfish, Bagridae  140
Yellowfish, Cyprinidae  139

## Invertebrates

African Clouded Yellow or Lucerne Butterfly  *Colias electo*  103, 136
African Honey Bee  *Apis mellifera*  137
African Monarch  *Danaus chrysippus*  37, 102,
Beetle  *Cetoniinidae*  16, 92
Beetle  *Coleoptera*  65
Beetle  *Lycidae*  96, 97
Blackfly  *Simuliidae*  96, 97
Blister beetle  *Mylabris*  18, 101, 138
Brownveined White  *Beleonois aurota*  82
Butterfly  *Lycaenidae*  100, 105, 136
Butterfly  *Satyridae*  136
Cape Honey Bee  *Apis mellifera capensis*  137
Carpenter bee  *Xylocopa caffra*  137
Cave cricket  *Speleiacris tabulae*  136
Centipede  *Chilopoda*  64
Christmas Butterfly  *see* Citrus Swallowtail
Cicada  *Cicadidae*  101
Citrus Swallowtail  *Papilio demodocus*  100, 136
Common Striped Hawk Moth  *Hippotion eson*  7, 137
Convolvulous Hawk Moth  *Agrius convolvuli*  33
Cotton-stainer beetle  *Dysdercus*  17
Crab spider  *see* Flower spider
Flower or Crab spider  *Thomisidae*  50, 69, 138
Fly  *Tabanidae*  101, 132
Foam grasshopper  *Dictyophorus spumans*  92
Freshwater Crab  *Potamon perlatus*  21, 139
Froghopper  *Cercopidae*  65
Garden Acraea  *Acraea horta*  57, 136
Ladybird  *Cheilomenes*  65
Leaf-mimicking grasshopper  *Truxalinae*  15, 92
Longhorned beetle  *Polyphaga*  59
Lucerne Butterfly  *see* African Clouded Yellow
Lycaenid butterfly  *see* Butterfly
Mayfly  *Ephemoptera*  21, 139
Mite  *Dinogamasus*  137
Moth  *Lepidoptera*  75, 78, 79, 136, 137
Mud wasp  *Sceliphron spirifex*  137
*Mulvia albizona*  100
Orb-web spider  *Argiope*  26, 32, 138
Paper wasp  *Vespidae*  60, 100
*Peripatopsis alba*  136
Pine-tree Emperor Moth  *Imbrassia cytherea*  31, 137, 149
Praying mantid  *Mantodea*  68, 92
Protea beetle  *Trichostheta fascicularis*  16
Pride of Table Mountain  *Meneris tulbaghia*  119
Pugnacious Ant  *Anoploplepis custodiens*  113, 137
Rain Spider  *Palystes natalius*  14, 137
Satyrid butterfly  *see* Butterfly
Scorpion  *Scorpiones*  138
Shield bug  *Pentatomidae*  92
Shorthorned grasshopper  *Acrididae*  48, 92
Silver Brown  *Pseudonympha magus*  100
Silver vlei-spider  *Metidae*  14
Social wasp  *Belonogaster*  60
*Spelaeogryphus lepidops*  136
Spottedwinged Antlion  *Palpares speciosus*  100
Stick insect  *Phasmatidae*  78, 92
Stonefly  *Plecoptera*  139
Tabanid fly  *see* Fly
Table Mountain beetle  *Colophon westwoodi*  136
Threadwaist wasp  *Sphecidae*  100
Tick  *Ixodoidea*  92
Tiger beetle  *Cicindela*  64, 138
Tortoise beetle  *Aspidomorpha*  101

Twig-wilter  *Anoplocnemis*  60, 61
Wasp  *Pompilidae*  137
Water beetle  *Dytiscus*  39, 138
Weevil  *Cuculonidae*  15, 100

## Alien Plants and Animals

*Acacia*  xi
Argentine Ant  *Iridomyrmex humilis*  142, 151
Australian Myrtle  *Leptospermum laevigatum*  (Australiese Mirt)  147, 153
Bass  *Micropterus*  (Baars)  150
Bibron's Gecko  *Pachydactylus bibronii*  150
Blackbird  *Turdus merula*  149
Black Wattle  *Acacia mearnsii*  (Swartwattel)  147, 148, 153
Bluegill Sunfish  *Lepomis macrochirus*  (Blouwang-sonvis)  150
Bramble  *Rubus*  (Braambos)  102
Brown Rat  *Rattus norvegicus*  (Bruinrot)  149
Brown Trout  *Salmo trutta*  (Bruinforel)  150
Carp  *Cyprinus*  (Karp)  150
Chaffinch  *Fringilla coelebs*  (Gryskoppie)  149, 150
Cluster Pine  *Pinus pinaster*  (Sparden)  147
Common Carp  *Cyprinus carpio*  (Karp)  150
*Cryptotermes brevis*  150
Eucalypt  *Eucalyptus*  153
European Fallow Deer  *Cervus dama*  (Europese Takbok)  149
European Starling  *Sturnus vulgaris*  (Europese Spreeu)  149
Feral Pig  *see* Wild Boar
Feral Pigeon  *Columba livia*  (Tuinduif)  150
Flowerpot Snake  *Ramphotyphlops braminus*  150
German Wasp  *Vespula germanica*  150
Goldfish  *Carassius auratius*  150
Grey Squirrel  *Sciurus carolinensis*  (Gryseekhoring)  149
Hakea  *Hakea*  147, 148, 151
Hakea beetle  *Melanterus servulus*  152, 153,
Hakea moth  *Carposina autologa*  152
Hakea snout beetle  *Erytenna consputa*  152
Hakea weevil  *Cydmaea binotata*  152
Helmeted Guineafowl  *Numida meleagris*  (Gewone Tarentaal)  150
Himalayan Tahr  *Hemitragus jemlahicus*  (Himalaja Tahr)  149, 154
House Mouse  *Mus musculus*  (Huismuis)  149
House Rat  *Rattus rattus*  (Huisrot)  149
House Sparrow  *Passer domesticus*  (Huismossie)  150
Longleaved Wattle  *Acacia longifolia*  (Langblaarwattel)  76, 147, 151, 152
Longleaved Wattle gall wasp  *Trichilogaster acaciaelongifoliae*  76, 152
Lucerne  *Medicago sativa*  103, 136, 154
Mallard  *Anas platyrhynchos*  150
Mute Swan  *Cygnus olor*  (Swaan)  150
Nightingale  *Luscinia megarynchos*  149
Oak  *Quercus*  (Eike)  149
Oleander  *Nerium oleander*  (Selonsroos)  153
Parrot's Feather  *Myriophyllum aquaticum*  (Waterduisendblaar)  147
Pine  *Pinus*  147, 148, 149, 151, 153
Port Jackson  *Acacia saligna*  (Port Jackson)  147, 151, 153
Rainbow Trout  *Salmo gairdenerii*  (Reënboogforel)  150
Rooikrans  *Acacia cyclops*  (Rooikrans)  147, 148, 149, 151, 153
Rook  *Corvus frugilegus*  149
Sesbania  *Sesbania punicea*  (Sesbania)  147, 152
Silky Hakea  *Hakea sericea = tenuifolia*  (Syerige Hakea)  103, 147, 148, 151, 152
Smallmouthed Bass  *Micropterus dolomieu*  (Kleinbekbaars)  150
Song Thrush  *Turdus philomelos*  149
St John's Wort  *Hypericum perforatum*  (Johanneskruid)  102, 147, 152
Stinkbean  *Albizia lophantha*  (Stinkboon)  152
Termite  *see Cryptotermes brevis*
Trout  *Salmo*  (Forel)  150
Wattle  *Acacia*  (Wattel)  147, 151, 153
Wild Boar or Feral Pug  *Sus scrofa*  (Wilde Huisvark)  149

162